FOLDASES CATALYZING THE FORMATION AND ISOMERIZATION OF DISULFIDE BONDS IN PROTEINS

Foldases Catalyzing the Formation and Isomerization of Disulfide Bonds in Proteins

Natalya K. Nagradova

Nova Biomedical Books
New York

Copyright © 2009 by Nova Science Publishers, Inc.

All rights reserved. No part of this book may be reproduced, stored in a retrieval system or transmitted in any form or by any means: electronic, electrostatic, magnetic, tape, mechanical photocopying, recording or otherwise without the written permission of the Publisher.

For permission to use material from this book please contact us:
Telephone 631-231-7269; Fax 631-231-8175
Web Site: http://www.novapublishers.com

NOTICE TO THE READER

The Publisher has taken reasonable care in the preparation of this book, but makes no expressed or implied warranty of any kind and assumes no responsibility for any errors or omissions. No liability is assumed for incidental or consequential damages in connection with or arising out of information contained in this book. The Publisher shall not be liable for any special, consequential, or exemplary damages resulting, in whole or in part, from the readers' use of, or reliance upon, this material.

Independent verification should be sought for any data, advice or recommendations contained in this book. In addition, no responsibility is assumed by the publisher for any injury and/or damage to persons or property arising from any methods, products, instructions, ideas or otherwise contained in this publication.

This publication is designed to provide accurate and authoritative information with regard to the subject matter covered herein. It is sold with the clear understanding that the Publisher is not engaged in rendering legal or any other professional services. If legal or any other expert assistance is required, the services of a competent person should be sought. FROM A DECLARATION OF PARTICIPANTS JOINTLY ADOPTED BY A COMMITTEE OF THE AMERICAN BAR ASSOCIATION AND A COMMITTEE OF PUBLISHERS.

Library of Congress Cataloging-in-Publication Data

ISBN: 978-1-60692-466-2

Available upon request

Published by Nova Science Publishers, Inc. ✦ New York

Contents

Preface		vii
Introduction		1
Chapter 1	Eukaryotic Protein Disulfide Isomerase (PDI)	3
Chapter 2	Disulfide Bond Formation and Isomerization in Prokaryotes	29
Conclusion		57
References		61
Index		69

Preface

One of the rate-limiting steps in the folding pathways of many secretory proteins is the formation of correct disulfide bonds between cysteine residues. In eukaryotes, both disulfide bond formation and isomerization which shuffles incorrectly formed disulfides are catalyzed by protein disulfide isomerase (PDI), whereas in bacteria these two reactions are catalyzed by separate enzymes. Both in eukaryotic and prokaryotic cells the oxidation and isomerization steps proceed exclusively in extracytoplasmic environments (the lumen of the eukaryotic endoplasmic reticulum and the Gram-negative bacterial periplasmic space). The family of foldases under discussion is characterized by a conserved "thioredoxin fold" and a common active site motif: Cys-X-X-Cys. The process of disulfide bond formation relies on thiol-disulfide exchange between oxidized and reduced cysteine pairs in the catalyst and substrate protein. Two separate pathways involved in disulfide bond formation and isomerization have been characterized both in eukaryotes and in bacteria. In the oxidative pathway, oxidizing equivalents flow from oxygen to a membrane protein (Ero1p in eukaryotes or DsbB in bacteria), and then to a folding protein containing reduced cysteines via PDI (in eukaryotes) or via DsbA (in bacteria). In the isomerization pathway, DsbC (bacterial protein disulfide isomerase) or PDI (in eukaryotes) interacts with substrate proteins that contain non-native disulfide bonds, allowing these bonds to rearrange to their native pairings. Reducing equivalents which are necessary to maintain DsbC in a reduced form, able to attack misfolded disulfides, are transferred from the cytoplasm with the aid of the cytoplasmic membrane protein DsbD. In eukaryotes, reduced glutathione is the main source of reducing equivalents for PDI. A dual role of PDI as an oxidase and an isomerase is facilitated by its complex domain architecture.

Introduction

Enzymes that are able to accelerate protein folding by catalyzing the rate-limiting steps of formation and (or) isomerization of disulfide bonds deserve special attention since one of them, namely eukaryotic protein disulfide isomerase, was the first ever to be identified as a foldase. Its discovery dates back to the early 1960s, when Christian Anfinsen and co-workers performed the classic experiment that demonstrated that oxidizing conditions are sufficient to refold reduced and denatured RNase A correctly [1]. It was observed that even though a correct formation of four disulfide bonds stabilizing the native structure of RNase A could occur spontaneously, there was an obvious discrepancy between the rate of refolding *in vitro* (hours to days) and *in vivo* (seconds to minutes), suggesting the presence of an *in vivo* catalyst for disulfide bond formation. The search for a potential cellular catalyst of this process resulted in the discovery of the first of such catalysts, protein disulfide isomerase (PDI), in the eukaryotic endoplasmic reticulum [2].

Over the next 40 years, it was found that PDI constitutes a major component of the lumen of endoplasmic reticulum in those cells that actively synthesize disulfide-bonded cell-surface and secretory proteins, and that changes in PDI abundance correlate with changes in the level of such synthesis [3-10]. A role for PDI in the catalysis of native disulfide bond formation in the endoplasmic reticulum was first established by mutational analysis in yeast, where the *PDI* gene was shown to be essential for cell viability and for oxidative protein folding [11, 12]. It was demonstrated that in intact cells, PDI can be cross-linked to nascent immunoglobulin chains, and that in a cell-free translation/translocation system the formation of disulfides in newly synthesized and translocated proteins depends on the presence of lumenal PDI [8]. Considerable advances have been made to date in elucidating the mechanisms of PDI-catalyzed reactions, and the recently determined crystal structure of this multidomain protein [13] has contributed significantly to the success of these studies.

In 1991, it was discovered that disulfide bond formation in the prokaryotic periplasm is a catalyzed process as well [14]. The availability of genetic tools in prokaryotes and the ease of handling of the proteins involved has allowed for rapid advances in understanding the process of disulfide bond formation and isomerization in bacterial periplasm, which yielded a substantial amount of new data. In this chapter, we shall review the current status of research devoted to foldases of this type, focusing on their structural properties and catalytic mechanisms. The role of protein-disulfide isomerases as molecular chaperones will be considered in detail in Chapter 5.

Chapter 1

Eukaryotic Protein Disulfide Isomerase (PDI)

1.1. The Role of Protein Disulfide Isomerase in the Endoplasmic Reticulum

PDI is a 55 kDa multifunctional protein located in the lumen of the endoplasmic reticulum where it is present at concentrations of approximately 2mM. According to its molecular mass, determined by size-exclusion chromatography, PDI has long been considered a homodimeric molecule [15, 16]. Recently analytical ultracentrifugation has shown that PDI is actually an elongated-shape monomer [17]. PDI is directed into the endoplasmic reticulum by a signal sequence and retained within the lumen by a recognition system for its C-terminal tetrapeptide motif (–KDEL in mammals or –HDEL in yeast) [8]. Those proteins destined for secretion are co-translationally translocated into the oxidizing environment of the endoplasmic reticulum ($E^{o'} = -180mV$), where they fold and acquire their native disulfide bonds. The redox state of the endoplasmic reticulum is considerably more oxidizing than that of the cytosol ($E^{o'} = -230mV$), a compartment where stable disulfide bond formation rarely occurs [18]. The oxidizing redox state of the endoplasmic reticulum is reflected in the relatively high intralumenal concentration of oxidized glutathione; the (GSH) : (GSSG) ratio has been estimated at about 3:1, the overall concentration of glutathione being in the millimolar range [19]. Therefore, a net influx of oxidizing equivalents is needed to support the rapid transit of secretory proteins through the endoplasmic reticulum, and to maintain a high concentration of GSSG within this organelle [19, 20]. That oxidizing function is fulfilled by **E**ndoplasmic **r**eticulum **o**xidoreductin 1 **p**rotein (Ero1p).

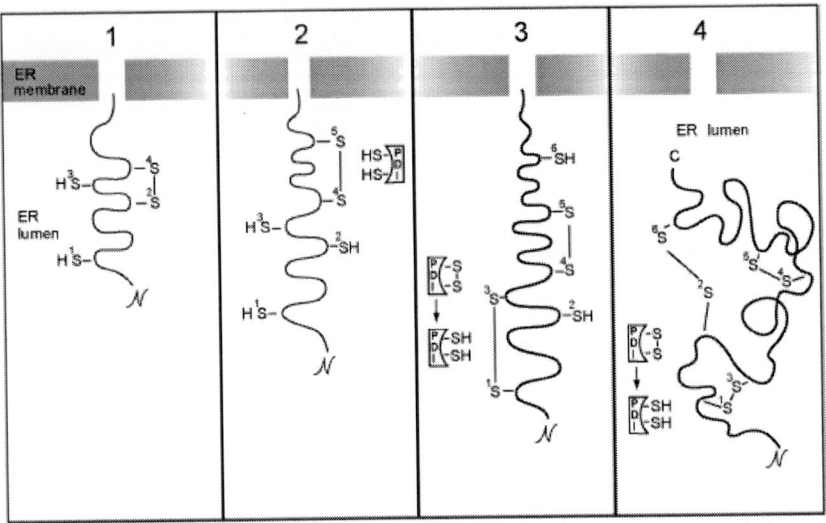

Figure 1. Schematic representation of the different steps of PDI-assisted folding of a nascent polypeptide chain translocated into the lumen of endoplasmic reticulum. The cysteine residues which form disulfide bonds in the native protein are numbered with superscripts above their S atoms. The numbers of these cysteines in the amino acid sequence of the protein are not indicated. Three co-translational (1-3) and one post-translational (4) stages are shown. In step 1, a non-native disulfide bond is formed spontaneously. Owing to the isomerase activity of PDI, a conformational rearrangement of the growing polypeptide chain results in the formation of a native disulfide bond (step 2). It is shown that PDI functions in its reduced form. The formation of a second disulfide bond is catalyzed by PDI operating as an oxidase (step 3). The final step occurs post-translationally; it includes the formation of a third correct disulfide bond, which stabilizes the N- and C-tails of the protein in their native conformation. Here PDI again functions as an oxidase. The N and C-termini of the polypeptide chain are indicated. ER= endoplasmic reticulum.

PDI catalyzes disulfide formation (oxidase activity) as well as the rearrangement of incorrect disulfide pairings (isomerase activity) [21, 22], accelerating both processes without drastically altering the refolding pathway [23-25]. The activity of PDI depends on a pair of cysteines located in the enzyme's active center. When the active site cysteines are present in disulfide form, the enzyme can transfer disulfide bonds directly to a pair of sulfhydryls in a substrate polypeptide, exhibiting an oxidase function. On the other hand, if the active site cysteines are present in dithiol form, the enzyme is able to catalyze disulfide reshuffling, serving as a disulfide isomerase. Figure 1 above schematically illustrates these functions of PDI at different stages of co-translational and post-translational folding of a nascent polypeptide chain entering the endoplasmic reticulum.

1.2. Modular Organization of the PDI Molecule and the Role of Different Domains

Structural information on PDI was first obtained by Edman et al., who determined the sequence for the rat enzyme [26]. These data supported the earlier proposal that PDI action is dependent on vicinal dithiol groups [27, 28]. It was shown that PDI contains two regions (the a and a' domains) closely similar in their amino acid sequences to that of the small protein thioredoxin [29], each domain containing a Cys-Gly-His-Cys active site. Further sequence analysis suggested two other potential domains, b and b', to be internal sequence repeats [26]. The order of these segments in the primary structure is a-b-b'-a'. The domain structure of PDI has been determined experimentally through a combination of protein engineering and limited proteolysis studies [30, 31]. Domain boundaries have been determined, and each domain has been expressed as an individual soluble folded protein. The domain structure (see Figure 2) confirmed the original domain assignments of Edman et al. [26], although with different domain limits [32].

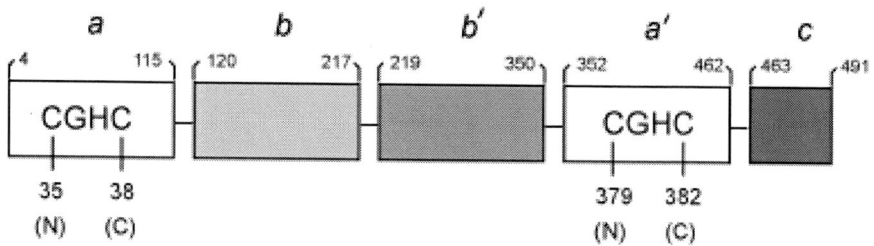

Figure 2. Domain structure of mammalian PDI (Darby et al. [30], Kemmink et al. [32], Noiva [21]). The first and last residues of each domain are indicated. The a and a' domains contain active sites comprising the CGHC motifs. The N-terminal cysteines (N) in these motifs are Cys35 and Cys379, whereas the C-terminal (C) cysteines are Cys 38 and Cys382, respectively. The C-domain comprises a 24 residue acidic segment in which over half the residues are glutamate or aspartate, followed by a KDEL sequence for retention in the endoplasmic reticulum.

The C-terminus of the polypeptide contains a highly acidic 29-amino acid extension c, which is not critical for any of the protein's major functions except ER retrieval [33].

The homology of amino acid sequences of the a and a' domains to thioredoxin made it likely that they would adopt a thioredoxin-like structural motif; this was indeed confirmed for the PDI a domain in a study by Kemmink et al., who determined the global fold of the human PDI a domain using heteronuclear multidimensional NMR [34]. As expected, the structure (comprising amino acids 1-120 of a total of 491 residues in the polypeptide chain) was shown to follow the thioredoxin motif, which is present in a number of related proteins [35]. The PDI a domain shares other notable features with E. coli thioredoxin; in particular, the two active site cysteine residues are in the same position at the N-terminus of helix α_2. As will be discussed below, this feature is functionally important.

Both the a and a' domains were shown to be capable of exhibiting similar and independent activities within PDI [36]. Yet when isolated, they only catalyzed a subset of the reactions involved in the forming of protein disulfide bonds [37], while proving inefficient for the the catalysis of disulfide rearrangements. This suggested that other domains, b and b' are needed for the isomerase function of PDI. To understand the role of these domains, a series of studies was undertaken. First of all, the structure of the b domain was determined [32]. It was found that, just like the a domain, the b domain also contains the thioredoxin motif. This observation, together with indications that the b' domain adopts a similar fold, implied that PDI should consist of active and inactive thioredoxin modules. It is likely that all four of the PDI modules arose by partial gene duplication or shuffling of a common thioredoxin gene, followed by divergence [38].

In the course of evolution, the b and b' modules appear to have lost the thioredoxin function of catalyzing thiol-disulfide exchange reactions. Neither module contains the two cysteine residues normally found at the active site of thioredoxin. Moreover, other residues normally contributing to the catalytic activity of thioredoxin, such as Asp26, Trp31 and Pro75 of human thioredoxin, which are conserved in the a and a' modules, have all been replaced in the b and b' modules [32]. All this is in line with the suggestion that the thioredoxin fold in the b and b' modules has been adapted to new functions, such as the binding of substrate proteins [32].

To clarify the nature of the interaction between PDI and its substrates, Klappa et al. [39] used chemical cross-linkers which can be applied to small amounts of proteins even in crude cell extracts. To characterize the substrate-binding site, they investigated the interaction of various recombinant fragments of human PDI expressed in *E. coli* with different radiolabelled model peptides. It was found that the b' domain of the enzyme is essential and sufficient for the binding of small peptides. Characterization of the primary substrate-binding site within the b' domain, performed by Pirneskovski et al., showed that the proposed ligand binding site is a small hydrophobic pocket defined by residues Leu242, Leu244, Phe258, and Ile272. Mutations within this site, expressed both in an isolated domain and in full-length PDI, greatly reduce the binding affinity for small peptide substrates [40].

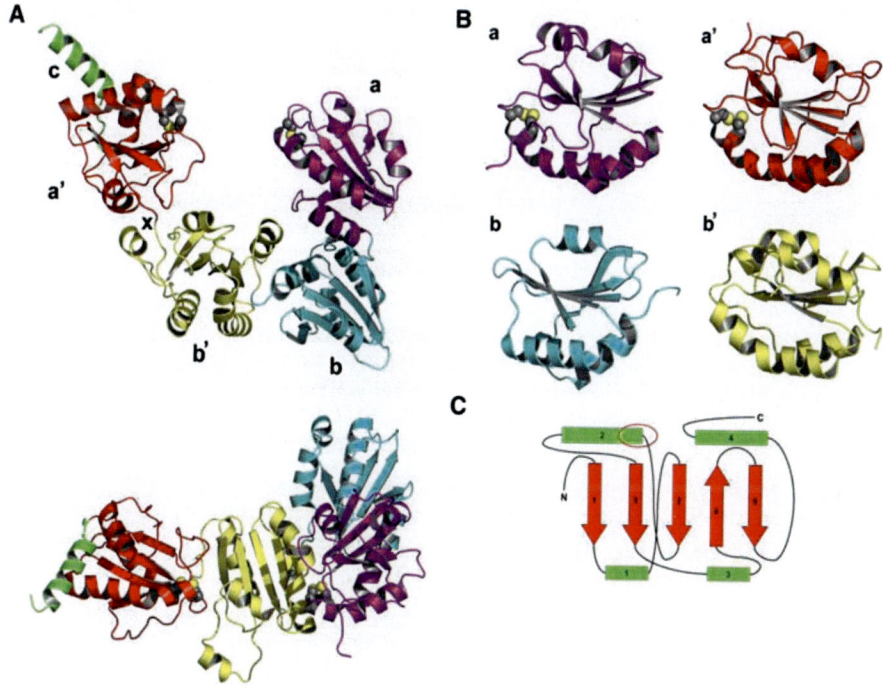

Figure 3. Overall structure of yeast PDI. A. Ribbon diagram of PDI with a, b, b' and a' domains in magenta, cyan, yellow, and red, respectively, and the C-terminal extension in green. The two orientations roughly differ by a 90° rotation around the horizontal axis. The side-chains of the active site cysteines in the a and a' domains are shown in space-filling representation with the sulfur atoms in yellow. B. Structural comparison of the individual domains of PDI. The domains are shown in the same native orientation, with the "long helix" side below the β sheet. The active site cysteine residues in the a and a' domains are shown in space-filling representation. C. Secondary structure diagram of the canonical thioredoxin fold with α helices in green and β strands in red. The location of the active site is indicated by a red oval. Reprinted from Tian et al. [13], with permission.

Contrary to the case of short peptides, the b' domain was shown to be essential but not sufficient for effective folding of larger peptides, specifically a 28-amino acid fragment derived from bovine pancreatic trypsin inhibitor, or of misfolded proteins, indicating that contributions from additional domains are required [39]. From the results obtained, a model has been proposed in which various domains of PDI contribute to the substrate binding. While the b' domain forms the essential core of the binding site and is sufficient for the binding of small peptides, the addition of further domains, such as a and a', provides for additional binding interactions and is therefore essential for the binding of more complex substrates. Finally, all of the domains might be required for a binding site that is large enough to interact extensively with an unfolded protein [39].

Such a model is fully consistent with the data obtained by Darby et al. in a parallel study aimed at characterizing the catalytic activities of PDI fragments composed of different domain combinations [41]. Thus, it was established that multidomain fragments of the PDI molecule have enhanced catalytic activities compared with individual a and a' domains, and that the full multidomain structure of the enzyme is particularly important for the catalysis of the complex disulfide rearrangements involved in the isomerase reaction. In sum, the whole body of information accumulated in the above studies strongly suggested that the efficient functioning of PDI as a protein folding catalyst should be predicated on the existence of an extensive substrate binding surface which might be assembled from the binding sites of individual domains combining to form a large substrate binding pocket [39, 41].

The three-dimensional structure of the entire PDI molecule, determined eight years later [13], proved to be fully consistent with the suggestions considered above. As shown in Figure 3 above, the crystal structure revealed that the four thioredoxin domains of the PDI molecule are arranged in a twisted U-shape, with the a and a' domains at the ends of the "U" and the b and b' domains forming the base. Notably, the inside surface of the "U" is rich in hydrophobic residues, facilitating interaction with misfolded proteins. The interface between the b and b' domains is rather extensive, with a buried surface area of ~700 $Å^2$. The crystal structure also showed that not only the b' domain, but also the b domain has an exposed hydrophobic patch at the same relative position. This, together with the hydrophobic patches surrounding the active sites in the a and a' domains, results in the formation of a continuous hydrophobic surface, which is crucial for the interaction between PDI and its substrates [13].

1.3. Functional Properties of PDI Active Centers. The pK$_a$ Values of Essential Cysteine Residues

The two pairs of cysteine residues in PDI, occupying positions which correspond to those of the essential dithiol/disulfide couple in thioredoxin, were suggested to constitute the active site groups of PDI [42]. To substantiate this suggestion, the properties of these cysteine residues were characterized in detail. The existence of two essential cysteines, Cys35 and Cys379, whose exclusive modification leads to inactivation of the enzyme, was detected in mammalian PDI. These groups were shown to possess unusually low pK$_a$ values equal to 6.7 and exhibited high nucleophilic reactivity in the fully ionized state [42]. As shown in Figure 2, these cysteine residues, designated N, are at the beginning of each of the active site motifs. A consideration of these results in the light of comparable data for thioredoxin [43] led the authors to conclude that the properties of the two enzymes are closely similar. The data obtained with PDI were interpreted to indicate that two disulfide bonds are located in thioredoxin-like domains at positions analogous to thioredoxin disulfide. When these disulfides are reduced to the dithiol state, the N-terminal cysteine residue of each pair has an unusually low pK$_a$, and hence is predominantly in the thiolate (-S$^-$) state at pH 7.5. Modification of these residues leads to complete inactivation. The second, C-terminal cysteine residue in each active site motif should have a high pK$_a$ (like thioredoxin whose C-terminal cysteine has a pK$_a$ of 8.7) [42 - 44].

The determination of the crystal structure of PDI has revealed that, as in all members of the thioredoxin family, the N-terminal cysteine of each PDI active site is located near the N-terminus of the second α-helix, which presumably increases its nucleophilicity [13]. Such a mechanism of lowering the pK$_a$ of the N-terminal cysteine had been suggested previously by Kortemme and Creighton [45]. It was shown clearly that the presence of a helical structure substantially influences the ionization properties of cysteine thiol groups located at the termini of model α-helices. Thiol ionization was favored at the N terminus and disfavored at the C-terminus, consistent with an interaction of the cysteine thiolate anion with the electrostatic field of the α-helix [45].

The pK$_a$ values of both the N-terminal and C-terminal active center cysteine residues play an important role in determining the physiological function of PDI. To act as an efficient thiol-disulfide oxidant (see Figure 4), the thiolate state of the N-terminal cysteine must be stabilized, and the thiolate state of the C-terminal cysteine destabilized. While increasing the pK$_a$ value of the C-terminal cysteine residue promotes the oxidation of substrates, it has an inhibitory effect on the reoxidation of the enzyme, since the latter process requires the formation of a thiolate at this position. Since reoxidation is essential to complete the catalytic cycle, a question arises of how the necessary change in the pK_a value of the C-terminal cysteine can be achieved. In search of a solution to this problem, Lappi et al. [46] have identified a conserved arginine residue located in the loop between β5 and α4 of the PDI catalytic domain (see Figure 4).

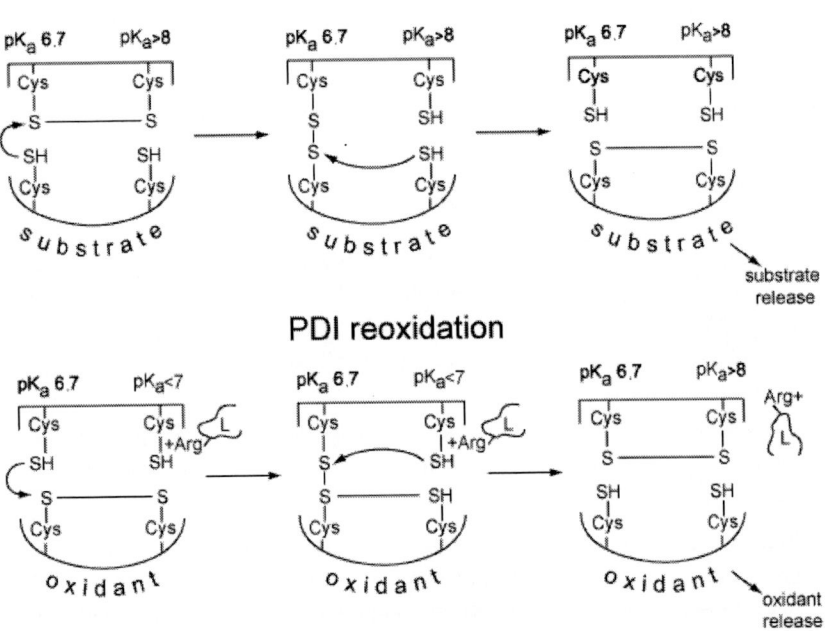

Figure 4. Schematic representation of the PDI-catalyzed substrate oxidation reaction (top) and the reaction of DPI reoxidation by an oxidant (bottom). Both reactions start with a PDI-substrate (or oxidant) mixed disulfide and proceed *via* thiolate formation on the substrate (top) or on PDI (bottom). In the latter case, the process is triggered by a change in the pK_a value of the C-terminal Cys residue in the PDI active center from a rather high value of more than 8 to a pK_a of <7, owing to the incorporation of a positively charged arginine residue into the active center. As noted in the text, this arginine belongs to a flexible loop connecting the β5 strand to the α4 helix in the structure of the a domain. As shown in the figure, thiolate formation is followed by a nucleophilic attack by this thiolate on the mixed disulfide. The pK_a values of the N- and C- cysteine residues are indicated. L designates the flexible loop comprising Arg120. Adapted from Lappi et al. [46].

This residue (R120 in the human PDI) turned out to be critical for the catalytic function of PDI, specifically for the reoxidation of the enzyme. An examination of the PDI a-domain structure, combined with molecular dynamic studies, suggested that the side-chain of this arginine residue can move into and out of the active site locale and that this has a very marked effect on the pK_a value of the active site cysteine residues [46]. For most of the time that the side-chain of R120 stays far away from the active site, the pK_a value of the C-terminal cysteine is high (consistent with experimental evidence), and the dithiol form of the protein is stabilized, allowing PDI to act as an oxidant. However, the side-chain of R120 has the potential to move closer to the active site, whereupon the pK_a value of the C-terminal cysteine falls dramatically, and the thiolate thus produced can act as a nucleophile on the mixed disulfide formed with an oxidizing agent (see Figure 4). The entry of the side-chain of R120 into the catalytic site offers an elegant solution to the problem of maintaining a high pK_a value for the C-terminal cysteine while allowing reactions which require a thiolate to proceed at an appreciable rate [46].

Recently another factor was identified that is able to influence the redox state of cysteine residues in the PDI active site. It has to do with the fact that, in addition to its four catalytic cysteines, PDI possesses two non-active site cysteines whose location and distance from each other varies from organism to organism. In mammals and birds, the non-active site cysteines are located in the b' domain of the protein, separated by 30 amino acids. However, in yeast and fungi, the non-active site cysteines are located near the N-terminal active site and are separated by just 6 to 8 residues. Until the last few years, the function of these cysteines and the significance of their unique location in yeast PDI remained unclear. In 2004, Xiao et al. showed that these non-active site cysteines are present *in vivo* as a stable disulfide [47] and probably serve to destabilize the N-terminal active site disulfide of the yeast PDI, making it a better oxidant [48].

1.4. Disulfide Isomerization: The Essential Function of PDI

1.4.1. The Most Significant Cellular Function of PDI Is the Isomerization of Non-Native Disulfide Bonds

An important function of PDI which does not involve any changes in the oxidation state of a substrate protein and requires neither a reduction nor a reoxidation of the enzyme, is the catalysis of protein disulfide bond isomerization. Abundant evidence suggests that disulfides that form at the early stages of protein folding often do so incorrectly; cysteines can be mispaired or disulfides can be formed in a wrong temporal sequence, making it difficult to oxidize buried cysteines [49-53]. To rectify these errors, incorrect disulfides must be broken down and new ones formed in a different configuration, and PDI has been shown to catalyze this process by rescuing kinetically trapped intermediates [25]. It has been established that PDI is essential for the viability of *Saccharomyces cerevisiae* [12, 54], and a considerable body of information supports the idea that of all the cellular roles of PDI, its most important function is the isomerization of non-native disulfide bonds [55-57, 12]. Thus, to determine why the *PDI* gene is essential for the growth of *S. cerevisiae*, Laboissiere et al. [56] mutated the cDNA that codes for PDI and tested the ability of the resulting mutant proteins to support the growth of *pdi1Δ S. cerevisiae* cells shown to be unviable [58]. The results obtained, coupled with *in vitro* analyses of catalysis, demonstrated that the essential role of PDI is not related to the net formation of protein disulfide bonds. Rather, its more important role is to act as a "shufflease", i.e. a catalyst of the isomerization of existing disulfide bonds [56].

1.4.2. The Mechanism of the Isomerization Reaction Catalyzed by PDI Depends on the Structure of Protein Substrate

Some peculiar characteristics of the isomerase reaction were revealed in the above study by Laboissiere et al. [56]. It was found that replacing PDI with a variant in which both active site sequences are CGHS instead of CGHC restores viability to *pdi1Δ S. cerevisiae*. Although this PDI variant has a low dithiol oxidation and disulfide reduction activity, it is proficient in its catalysis of disulfide isomerization [56]. On the other hand, a PDI variant that contains SGHC active-site sequences is unable to catalyze the formation, reduction or isomerization of disulfide bonds, and does not complement a *pdi1* deletion. These data suggested that the essential functional group in the CGHC motif required for PDI isomerase reaction is the sulfhydryl group of the N-terminal cysteine residue. Thus, the simplest mechanism for the catalysis of an isomerization reaction could be described as follows. The process begins with the attack of a thiolate ion on a protein disulfide, forming a mixed disulfide [37]. The protein thiolate thus produced can then attack another protein disulfide bond. Finally, the resulting thiolate attacks the mixed disulfide to release an unaltered catalyst. Such an isomerization reaction would be driven by the search for the most stable conformation of the substrate protein [56].

However, several years later Gilbert and co-workers [53, 59, 60] realized that the mechanism of PDI-catalyzed isomerization may be more complex. Under some conditions substrate thiolates proved unable to complete the required disulfide rearrangements, effectively trapping an enzyme-substrate mixed disulfide. In such a case, another cysteine residue of the PDI active center, namely the C-terminal cysteine residue of the CGHC motif, would come into play, triggering an "escape" mechanism shown in Figure 5. This cysteine residue forms a disulfide bond in the active site, releasing a reduced substrate. Cycles of reduction and reoxidation would ultimately produce a native protein [53].

Experimental evidence for such a mechanism was obtained in the study of a mutant containing only the N-terminal cysteine at the active site. Given that the disulfide rearrangement initiated by an attack of a single active-site cysteine could propagate through the substrate and terminate by expelling PDI, the authors were surprised to find that PDI mutants with only one (the N-terminal) cysteine at the active site were considerably less effective than a wild-type active site in catalyzing rearrangements of RNase A during oxidative folding of the reduced substrate [59]. It was found that mutant active sites of the CXXS type exhibited only 12%-50% activity in catalyzing disulfide rearrangement; the lower end of their activity range was also associated with an accumulation of covalent intermediates between PDI and its substrate. In sum, these data led the authors to propose a mechanism in which the second (C-terminal) active-site cysteine offers PDI a way to "escape" from covalent intermediates that do not rearrange in a timely fashion. Moreover, according to this hypothesis, the second active-site cysteine may serve as an internal clock in the wild-type enzyme, limiting the time allowed for intramolecular substrate rearrangements under normal conditions [59]. In subsequent years, Walker and Gilbert provided evidence for the assumption that escape through the second active site cysteine is in fact the dominant mechanism for PDI-dependent isomerization of scrambled RNase A [60].

The question, however, remained of whether the steps that follow the formation of the initial mixed disulfide are similar in all types of substrates that undergo PDI-catalyzed disulfide bond isomerization. Is the ability to "escape" a mechanistic imperative? To develop a more detailed analysis of the mechanism of isomerization, Kersteen et al. [61] used a homogenous substrate, the 17-mer peptide tachyplesin 1 that folds into a β - hairpin stabilized by two disulfide bonds. A variant of this peptide was synthesized in which two disulfide bonds

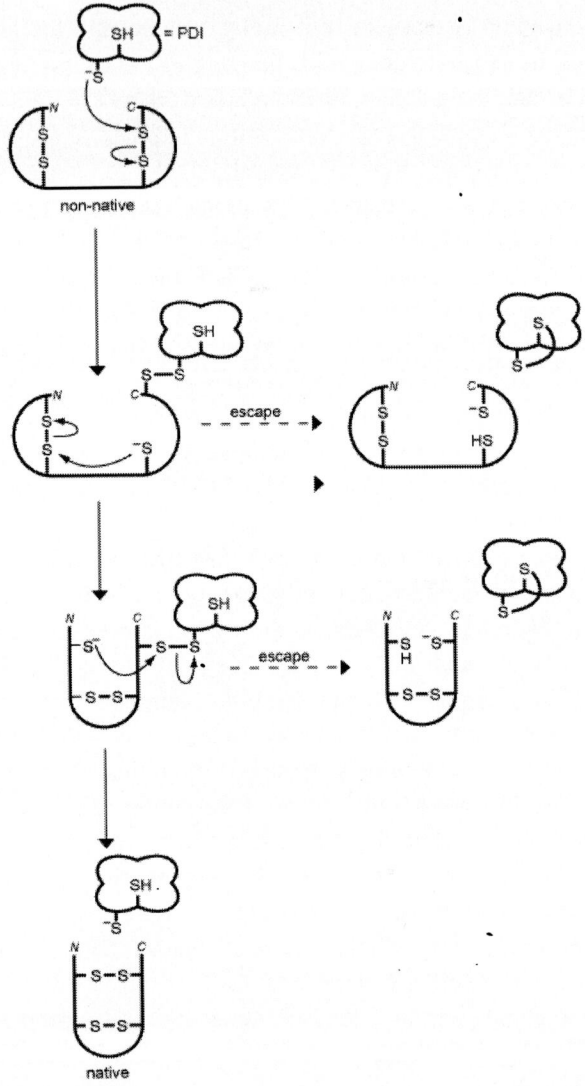

Figure 5. Putative mechanism of disulfide bond isomerization by PDI (Kersteen et al. [57], Woycechowsky et al. [55]). The reaction begins with a nucleophilic attack by a PDI-provided thiolate on a non-native disulfide bond. The resulting covalent substrate-catalyst complex contains a substrate thiolate and can perform intramolecular thiol-disulfide exchange reactions to ultimately produce native disulfide bonds and regenerate PDI. If a mixed disulfide intermediate becomes trapped, the catalyst escapes from the covalent complex, the escape being initiated by the second thiol in the active site of PDI. Subsequent oxidation of the substrate and reduction of the enzyme would be necessary to produce native disulfide bonds and regenerate PDI. See text for details.

were in a non-native state and side chains near its N and C-termini contained a fluorescent donor (tryptophan) and acceptor (N_ε-dansyllysine). Upon isomerization of the non-native disulfide bonds into their native state, a 28-fold increase in fluorescence resonance energy transfer from 280 to 465 nm took place. This continuous assay was used to analyze catalysis by wild-type human PDI and by a variant in which the C-terminal cysteine residue within each CGHC active site had been replaced with alanine.

It was found that wild-type PDI catalyzes the isomerization of this small model substrate with $k_{cat}/K_M = 1.7 \times 10^5$ M^{-1} s^{-1}, which is the highest value yet reported for catalysis of disulfide bond isomerization. The variant, which was a poor catalyst of disulfide bond reduction and dithiol oxidation, appeared to retain virtually all of the activity of wild-type PDI in catalyzing the isomerization of disulfide bonds. Thus, C-terminal cysteine residues do not play a significant role in the isomerization of the disulfide bonds in non-native tachyplesin 1, suggesting that catalysis of disulfide bond isomerization by PDI does not necessarily involve a cycle of substrate reduction/oxidation [61]. It is apparent that in the case of a small peptide, PDI is able to catalyze disulfide bond isomerization by a simpler mechanism than that used with a scrambled RNase A, which undergoes numerous reduction-reoxidation cycles. It therefore seems likely that the details of the mechanism may vary depending on the structural complexity of the protein substrate, and there may be no "universal" mechanism for catalysis of disulfide isomerization [53].

1.4.3. Relationship between the Overall Structure of PDI Molecule and Its Activity as a Protein Disulfide Isomerase

The overall structural similarity between PDI and DsbC, a periplasmic protein disulfide isomerase, is remarkable. As described below, DsbC forms a homodimer [62]; each subunit has an N-terminal dimerization module followed by a thioredoxin-like domain containing the active-site cysteines. Overall, the molecule has a V-shape, with the active sites located near the ends of the "V" and facing each other. The inside of the cavity between the monomers is rich in hydrophobic residues, facilitating interaction with unfolded non-native protein substrates. As seen in Figure 3, similar structural features are observed in PDI. A reasonable explanation of this similarity may be that it is the result of convergent evolution which created a common functional basis for protein disulfide isomerization. A particular arrangement of the four PDI domains into a single polypeptide chain forming a U-like structure, on the one hand, and an association of two monomeric

DsbC molecules into a V-shaped dimer, on the other hand, both result in structures characterized by (i) a close spatial proximity of two thioredoxin-like domains on the opposite sides of a large pocket, and (ii) extensive protein-binding surfaces rich in hydrophobic residues. Thus, it seems likely that these features may be essential for isomerase activity. This suggestion is supported by the fact that individual catalytic domains of human PDI have a very low isomerization activity of ~5% [41], while heterodimers of DsbC with one active site and one inactive thioredoxin domain are basically inactive [63]. As noted by Tian et al. [13], one possible explanation for this fact could be that PDI with two active thioredoxin domains, like the DsbC dimer, has the ability to more or less simultaneously reduce two incorrect disulfide bonds, which diminishes conformational strain in the substrate and facilitates the correct pairing of cysteine residues. This is in contrast to proteins with single thioredoxin domains, where an initially incorrect disulfide bond might simply reform after some time because the substrate is conformationally trapped by the remaining incorrect disulfide pairs [13]. The second common feature of PDI and DsbC is their ability to interact efficiently with incompletely folded or misfolded proteins, thereby exhibiting chaperone functions. The peptide binding sites present in these enzymes contribute significantly to the isomerase activity by selecting proteins with incorrect disulfide bonds via hydrophobic interactions, which increases their local concentrations [13].

1.4.4. Is PDI Capable of Discriminating between Native and Non-Native Disulfides?

As considered above, the mechanism of PDI-catalyzed disulfide isomerization involves cycles of disulfide reduction and oxidation, so that incorrect disulfides are reduced and eventually converted, through trial and error processes, to native disulfides that are resistant to further isomerization. To find out whether PDI can distinguish native structures from non-native ones, Zheng and Gilbert compared the reactivity of several native protein disulfides with dithiothreitol (DTT) in the presence and absence of PDI [64]. Three disulfide-containing native proteins: RNase A, microbial ribonuclease T1, and bovine pancreatic trypsin inhibitor, were examined for PDI-catalyzed reduction. The first-order dependence of PDI-catalyzed reduction on the concentration of DTT, substrate protein and PDI suggested a simple mechanism where the rate-limiting step would involve direct reduction of a disulfide formed as a result of an attack by PDI on the native protein. For native proteins, the rate constants for PDI-catalyzed reduction correlated very well with the rate constants for uncatalyzed reduction by DTT.

To compare PDI-catalyzed reduction of native disulfides with reduction of a random, unstructured disulfide, the reduction of scrambled RNase A was measured. In this case, the reduction was much faster: the rate constant for PDI-catalyzed reduction of scrambled RNase A was found to be about 60 times faster than expected from the rate of its reduction by DTT. This indicates that although PDI does not recognize differences between various native proteins any better than DTT, it does detect unfolded substrates better than DTT, presumably through interaction with an extremely unfolded polypeptide. This more extensive interaction would allow PDI to kinetically detect disulfides in the context of unfolded polypeptide and specifically reduce them. In such a manner, the chaperone activity of PDI contributes significantly to its catalytic function. This property, exhibited by different types of protein isomerases, will be further discussed in Chapter 5.

1.5. Dithiol Oxidation Catalyzed by PDI

In vitro assays for oxidative protein folding typically employ a small molecule, such as oxidized glutathione (GSSG), as an electron acceptor for the oxidation of protein dithiols to disulfide bonds. The rate of oxidative protein refolding, regardless of the presence of a catalyst, depends on the redox potential of the assay buffer. For example, the optimum rate of PDI-catalyzed refolding of reduced RNase A occurs at a (GSH) : (GSSG) ratio of 5:1 in the presence of 1 mM total glutathione [65]. Measurements of the redox conditions within the lumen of endoplasmic reticulum show that the most abundant redox buffer present is glutathione and that the (GSH): (GSSG) ratio is about 2:1. In contrast, the (GSH) : (GSSG) ratio in the cytosol ranges from 30:1 to 100:1 [19]. These findings point to a mechanism that concentrates GSSG in the endoplasmic reticulum lumen as a potential source of oxidizing equivalents for the formation of protein disulfide bonds [66, 67, 21]. However, until recent years, the source of oxidative equivalents utilized for disulfide bond formation in the endoplasmic reticulum was unclear.

In 1998, the Kaiser [66] and Weissman [67] groups independently discovered a protein which plays an essential role in oxidative protein folding; this protein was termed Ero1p (endoplasmic reticulum oxidoreductin 1p) [67]. A genetic dissection of oxidative protein folding in yeast began with the isolation of an essential and conserved gene, *ERO1*, encoding a novel endoplasmic reticulum membrane protein which is required for protein oxidation [66, 67]. A temperature-sensitive allele of *ERO1* was identified in a screen for mutants defective in the export from the endoplasmic reticulum of secretory proteins containing disulfide bonds. Secretory proteins that would normally acquire intramolecular disulfide bonds remained completely reduced in the conditional *ero-1* mutant [66]. A conclusion that Ero1p introduces oxidizing equivalents necessary for protein disulfide bond formation in the lumen of the endoplasmic reticulum was supported by the observation that a membrane-permeable thiol oxidant diamide can substitute for the *ERO1* function [66]. Moreover, overexpression of *ERO1* increased the oxidative capacity of the cell, whereas a loss of Ero1p function resulted in a dramatic induction of the unfolded protein response [66, 67].

1.5.1. Ero1p: An Enzyme Producing Disulfide Bonds for Oxidative Protein Folding in the Endoplasmic Reticulum

The C-Terminal Domain of Yeast Ero1p Mediates Membrane Localization and Is Essential for Function

In addition to yeast Ero1p, two human homologues, Ero1-Lα and Ero1-Lβ were found and characterized [68, 69]. Also in human cells, PDI forms mixed disulfides with Ero1-Lα [70] and Ero1-Lβ [71], suggesting that the pathways of disulfide bond formation are conserved between yeast and human cells [72]. Both human Ero1-Lα and Ero1-Lβ complement a yeast thermosensitive mutant strain *(ero1-1)*, indicating functional conservation among the Ero1 family. Indeed, the protein sequences are rather conserved, especially around the CXXCXXC motif, which is functionally important [66, 68, 70]. However, a typical feature of the yeast protein is its C-terminal tail of 127 residues, which is absent in the two human proteins [68].

In a comparative study performed with yeast Ero1p and human Ero1-Lα proteins, Pagani et al. [72] showed that while Ero1p is capable of tightly associating with the endoplasmic membrane, Ero1-Lα and a yeast mutant (Ero1pΔC) lacking the 127 C-terminal amino acids are soluble when expressed in yeast. Neither Ero1-Lα nor Ero1pΔC complements an *ERO1* disrupted strain. Appending the yeast C-terminal tail to human Ero1-Lα restores membrane association and allows growth of *ERO1* disrupted cells. This indicates that in yeast cells, membrane association of Ero1p mediated by its C-terminal tail is crucial for function.

1.5.2. The Evidence that Oxidizing Equivalents Flow from Ero1p to Substrate Proteins via PDI

The development of several analytical techniques enabled the group of Kaiser to assay the redox state of endoplasmic reticulum proteins and to capture intermolecular mixed disulfides between proteins undergoing thiol-disulfide exchange in this compartment. Using these methods, Frand and Kaiser [73] have explored the functional relationship between Ero1p, PDI, and the secretory marker protein carboxypeptidase Y (CPY). First, it was shown that PDI-depleted cells are defective in the net formation of protein disulfide bonds in CPY, indicating that PDI acts as an oxidase *in vivo*. The capture of mixed disulfides between PDI and p1 CPY (the newly synthesized CPY), performed by the authors, indicates that PDI engages directly in thiol-disulfide exchange with newly synthesized secretory proteins. In wild-type cells, the active site cysteines of PDI are oxidized, but become reduced in a conditional *ero1-1* mutant, suggesting that Ero1p is responsible for the oxidation of PDI.

Second, mixed disulfides between PDI and Ero1p were also detected, consistent with a direct transfer of oxidizing equivalents from Ero1p to PDI. Since Ero1p itself remains oxidized in PDI-depleted cells, it was concluded that oxidizing equivalents do not flow in the reverse direction, from PDI to Ero1p. These results defined a pathway for protein oxidation in the endoplasmic reticulum wherein PDI serves as an intermediate in the transfer of oxidizing equivalents from Ero1p to substrate proteins. Protein disulfide bond formation may proceed through this pathway without a requirement for oxidized glutathione [73].

1.5.3. The Mechanism of Ero1p Oxidative Activity in Protein Disulfide Bond Formation

The discovery that Ero1p engages directly in thiol-disulfide exchange with PDI predicted that at least one redox active disulfide bond would be required for the functioning of this protein. In an effort to identify the redox-active cysteines in Ero1p, Frand and Kaiser [74] performed a mutational analysis of the seven cysteine residues that are absolutely conserved in the eukaryotic sequence homologues of Ero1p. This analysis identified two pairs of conserved cysteines, namely Cys100-Cys105 and Cys352-Cys355 of yeast Ero1p, which are required for efficient oxidative protein folding in the endoplasmic reticulum. Substitution of Cys100 with alanine impedes the capture of a mixed disulfide complex with

PDI in wild-type cells, and also blocks oxidation of PDI *in vivo*. Substitution of Cys352 or Cys355 with alanine prevents reoxidation of Ero1p *in vivo*.

The observation that those cysteine residues, participating in efficient mixed-disulfide formation with PDI, also are required for oxidative protein folding in the endoplasmic reticulum, supported the model according to which thiol-disulfide exchange between Ero1p and PDI drives the major pathway for protein oxidation in eukaryotic cells [75]. Taken together, the results considered above suggested the following mechanism for electron transfer by Ero1p. Cys100 and Cys105 form a redox-active disulfide bond that preferentially engages in thiol-disulfide exchange with PDI. Cys352 and Cys355 form a second redox-active disulfide bond that serves to reoxidize the Cys100-Cys105 cysteine pair, possibly through an intramolecular thiol-disulfide exchange reaction. Reoxidation of Ero1p could proceed via the transfer of electrons from the Cys352-Cys355 cysteine pair to an electron acceptor which remained unknown at the time of those experiments [74].

Soon thereafter, the missing acceptor was identified by another research group: Tu et al. demonstrated that Ero1p-catalyzed disulfide formation is driven by a flavin-dependent reaction [76]. Depletion of riboflavin, a precursor of flavin adenine dinucleotide (FAD), rapidly causes defects in the folding of disulfide-containing substrates. Further evidence that oxidative folding in eukaryotes is dependent on FAD came from the discovery that purified Ero1p itself is a novel FAD-binding protein [76]. When FAD is added to purified Ero1p and PDI, these components together can ensure robust oxidative folding *in vitro*. Ero1p contains tightly associated FAD, as evidenced by the fact that up to 50 percent of Ero1p purified from yeast microsomes firmly retains bound FAD [76].

Thus, the overall reaction can be divided in two half-reactions. In the reductive half-reaction, the enzyme accepts electrons from reduced PDI, resulting in a reduction of the bound flavin cofactor. In the oxidative half-reaction, the enzyme deposits the electrons on the acceptor (molecular oxygen) to restore the bound cofactor to its initial state as shown in Eqs.1 and 2.

$$E \cdot FAD + PDI\text{-}(2SH) \rightarrow E \cdot FADH_2 + PDI\text{-}(S\text{-}S) \tag{1}$$

$$E \cdot FADH_2 + O_2 \rightarrow E \cdot FAD + H_2O_2 \tag{2}$$

It was suggested that Ero1p can use molecular oxygen as a terminal electron acceptor, so that Ero1p and PDI act as a "self-contained" oxidase system that directly couples disulfide formation to the consumption of molecular oxygen [76]. Using a combination of a genetic and a biochemical approach, Tu et al. obtained evidence for this suggestion. They demonstrated that although FAD is critical for sustained Ero1p activity, it does not function as the terminal electron acceptor. Each FAD-bound Ero1p molecule is able to support multiple rounds of PDI oxidation, and an excess of free FAD cannot drive Ero1p-catalyzed disulfide formation under anaerobic conditions [77]. These observations indicated that molecular oxygen rather than FAD serves as the terminal electron acceptor.

In vitro experiments confirmed that Ero1p-catalyzed disulfide formation is compromised under anaerobic conditions, and that Ero1p directly consumes molecular oxygen during its reaction cycle [77, 78]. The efficient use of molecular oxygen as the terminal electron acceptor by FAD-bound Ero1p could explain how the oxidation of millimolar concentrations of PDI [79] is achieved despite FAD concentrations in the low micromolar range [80]. Recently, Gross et al. presented compelling evidence that under aerobic conditions, reduction of molecular oxygen by Ero1p yields stoichiometric amounts of hydrogen peroxide. Remarkably, it was also found that reduced Ero1p can transfer electrons to a variety of small and macromolecular electron acceptors beside molecular oxygen [81]. In particular, Ero1p can catalyze the reduction of exogenous FAD in solution. Free FAD is not required for the catalysis of dithiol oxidation by Ero1p, but it is sufficient to drive disulfide bond formation under anaerobic conditions [81]. This reduction of free flavin by Ero1p could not be the result of an exchange between bound and free flavin because, under the conditions of the assay, no loss of the stoichiometrically bound FAD cofactor was detected [81]. The authors suggest that the flavins' ability to function as electron acceptors for Ero1p may be shared by other chromophores, though the identity of these acceptors remains unknown [78, 81].

1.5.4. The X-Ray Crystal Structure of Ero1p Reveals the Spatial Relationship between Functional Cysteines and the Bound FAD

To obtain a deeper insight into the structural basis of Ero1p functioning, Gross et al. solved the crystal structure of this protein [82]. Their main objective was to elucidate the relationship between the elements characteristic of Ero1p, namely bound flavin, and the two pairs of cysteines that accept electrons from PDI. The overall fold of Ero1p appeared to be novel, having no structural neighbors [82]. A single domain of this protein is predominantly α-helical, with five short β-sheets and two poorly structured extended loops. The FAD cofactor is held between helices α2 and α3, in a bent conformation with the isoalloxasine and adenine rings buried within the protein. As a result of this packing, the isoalloxasine ring system is presented to the active site cysteines: the disulfide formed between Cys352 and Cys 355 is the prime candidate for a catalytic disulfide [82, 83]. In accordance with the model suggested by Frand and Kaiser [74] (see above), it is reasonable to suspect that FAD acts directly to oxidize this pair of cysteines [83].

Further crystallographic evidence helped explain the mechanism of the intramolecular disulfide relay suggested by Frand and Kaiser, whereby the Cys100-Cys105 pair accepts electrons from PDI and tunnels them to the active-site disulfide Cys352-Cys355 [74]. In fact, a comparison of the Ero1p structures derived from the two crystal forms revealed a striking degree of flexibility in certain key regions. The loop containing the Cys100-Cys105 disulfide bond adopts two conformations differing from each other by as much as 17Å, with the cysteine sulfur atoms displaced by 5.5 Å. In one crystal form, Cys100-Cys105 disulfide is located immediately adjacent to the Cys352-Cys355 pair. In the other crystal form, the Cys100-Cys105 disulfide has swung out and becomes surface-exposed [82, 83]. A slight modification of the protein backbone in this region would position Cys105 within disulfide-bonding distance of Cys352, which is a requirement for dithiol-disulfide exchange. Thus, the flexibility in the Cys100-Cys105 region may solve the steric problem of how this loop can interact alternately with PDI and the Ero1p active site [82, 84]. Figure 6 schematically illustrates the participation of the flexible loop in the Ero1p catalytic mechanism.

1.5.5. Two Catalytic Centers of PDI Are Not Equivalent

Elucidation of the native pathway of electron flow from reduced PDI to Ero1p raised a question of whether both active sites of the PDI molecule are equally efficient in their oxidative activity. The problem of functional non-equivalence between PDI active sites had been addressed in a number of previous studies, which revealed some differences in the kinetic behavior of N-terminal and C-terminal PDI active sites [37, 85-89]. A detailed examination of the structural features of both the a and a' domains of yeast PDI showed that they are very similar to each other and can be superimposed with a root mean square deviation of 1.3 Å for 98 residues [13]. However, despite highly similar overall structure, a few significant differences between the active sites were identified, leading the authors to suggest that they may reflect different catalytic properties of these sites [13].

The most significant difference between the a and a' domains appeared to be the redox state of their active site cysteines. Whereas in the a domain the two cysteines are in preference in the oxidized state, forming a disulfide bridge, their counterparts in the a' domain are in the reduced state [13]. The standard redox potentials were determined to be -180mV for the a domain and -152mV for the a' domain, suggesting that the a' domain is a better oxidant that the a domain. These values predict that the a domain is more stable in the oxidized form; while the a' domain prefers to be in the reduced form, in good agreement with the X-ray structure [13]. It thus seems plausible that it is the a' site that is selectively oxidized by Ero1p in the course of interaction with its flexible loop (see Figure 6, 3). Experimental support for this suggestion was obtained by Tsai and Rapoport, who were able to demonstrate on a model system that Ero1p can selectively oxidize one of PDI's two disulfide bonds, and showed that bond to be the one located in the COOH-terminal thioredoxin domain [90].

Recently, the Weissman group carried out a systematic study aimed at exploring the role of individual PDI domains in disulfide bond formation during a reaction driven by their natural oxidant, Ero1p [91]. It was found that Ero1p oxidizes the isolated PDI catalytic thioredoxin domains, a and a', at the same rate. A discrepancy between the rates of oxidation of the two active sites is only observed in the context of the full-length protein, with the C-terminal site acting primarily as an oxidase while the N-terminal site acts as an isomerase. To shed light on the mechanism of such functional asymmetry a possibility was studied that substrate binding to PDI might play a role in altering the rate of Ero1p-mediated oxidation. It was assumed that the substrate can affect the rate of the reaction in two ways: first, as a source of reductant for PDI, and second, as an unfolded protein that binds to PDI and may alter the ability of Ero1p to interact with PDI. To focus specifically on the effects of substrate binding in the absence of catalysis, the authors measured the rate of oxidation of PDI in the presence of reduced and alkylated RNase A [91]. This ingenious approach allowed them to discover that substrate binding specifically protects the N-terminal thioredoxin domain from oxidation by Ero1p. Thus, the oxidation rate of the first a active site fell to 40% in the presence of alkylated RNase A. In contrast, the oxidation rate of the a' active site was not altered significantly in the presence of this substrate. Furthermore, the above effect was found to be dependent on the presence of an intact b' thioredoxin domain [91].

In this manner, substrate-mediated protection of the N-terminal active site magnifies the intrinsic difference in oxidation rates between PDI active sites and contributes further to the asymmetry in their functions. The resulting disparity in the rates of oxidation of the two active sites ensures that the a' domain is in the oxidized state, which is required to catalyze disulfide formation, while the a domain is reduced and can promote disulfide isomerization. The authors propose

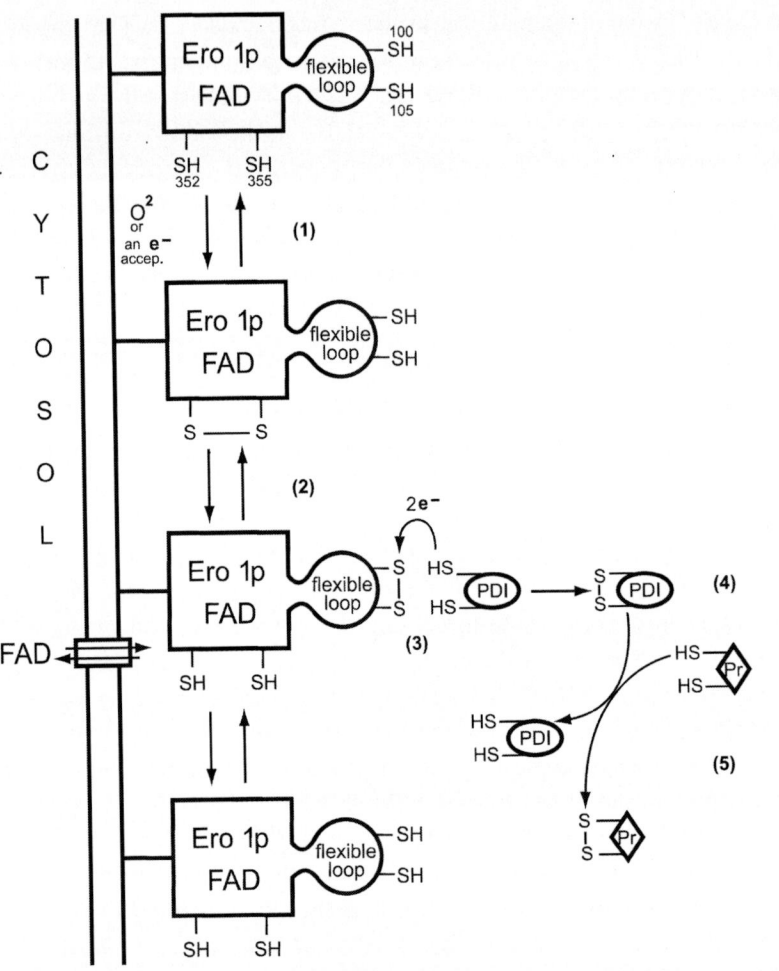

Figure 6. A model for Ero1p-mediated oxidative folding in the endoplasmic reticulum. The FAD cofactor is buried deep within Ero1p. Two pairs of cysteine residues are numbered according to their position in the yeast Ero1p sequence. The disulfide between Cys352 and Cys355 is adjacent to flavin. The loop containing Cys100 and Cys105 is flexible and may be involved in shuttling electrons from the substrate to the Cys352 - Cys355 disulfide. The FAD cofactor interacts with an oxygen molecule to generate a disulfide bond between nearby cysteines. Under anaerobic conditions, an alternate terminal electron acceptor can be used (1). The disulfide (Cys352-Cys335) is then probably transferred to the Cys100-Cys105 pair (2). This disulfide is located on a flexible loop of the protein which could move through the protein molecule and bring the disulfide into contact with PDI (3). After the disulfide is transferred to PDI (4), this enzyme is ready to oxidize a substrate protein (Pr), (5). Adapted from Gross et al. [81], Bardwell [83], and Tu et al. [77].

an interesting hypothesis to the effect that the regulation of PDI redox state by substrates may have important implications for oxidative protein folding in the endoplasmic reticulum. When there are fewer substrates to fold, Ero1p can keep PDI in a completely oxidized state, poised for the transfer of disulfides to substrate proteins at a maximal rate. Once PDI has engaged the substrate, it can then switch to its dual roles as an oxidase and an isomerase. It is also possible that the nature of the substrate might dictate the degree of protection for the N-terminal active site and thus modulate the roles of oxidation *versus* isomerization, depending on the requirements of the substrate [91].

1.6. The Role of Glutathione in Oxidative Protein Folding in Endoplasmic Reticulum

1.6.1. Oxidized Glutathione Does Not Provide the Oxidation Equivalents Necessary for the Formation of Disulfide Bonds

Glutathione is the major redox buffer in the endoplasmic reticulum, and the ratio of the concentration of GSH to GSSG (1:1 to 3:1) is similar to that found in redox buffers affording optimal rates for oxidative refolding *in vitro* [19]. This abundance of GSSG in the secretory pathway was long thought to be the primary source of oxidizing equivalents for disulfide bond formation *in vivo*, as it is *in vitro* [19]. Careful measurements could detect no other small-molecule redox component that was capable of cross-linking to a cysteine-containing peptide. Accordingly, disulfide bonds in newly synthesized proteins could be presumed to form through thiol-disulfide exchange reactions with GSSG or with oxidases dependent upon GSSG as a source of oxidizing equivalents [20, 92]. However, in the late 1990, experiments aimed at examining the function of glutathione in the endoplasmic reticulum *in vivo*, provided results that laid this hypothesis to rest. It was demonstrated that the major pathway for oxidation in the yeast endoplasmic reticulum, defined by the protein Ero1, is responsible for the oxidation of both glutathione and protein thiols, and that glutathione competes with protein thiols for the oxidizing machinery [66, 67, 93].

These studies relied in part upon a yeast mutant that is devoid of intracellular glutathione owing to a disruption of *GSH1*, the gene encoding the cytosolic enzyme which catalyzes the first, rate-limiting step in glutathione biosynthesis. Mutation of *GSH1* produces cells with no detectable intracellular glutathione [94]. GSSG was found to be entirely dispensable for the purposes of protein oxidation *in vivo* because oxidative protein folding would proceed with normal kinetics also in cells lacking glutathione [66]. A reasonable way to determine whether glutathione oxidation in the endoplasmic reticulum depends on the same pathway as that required for the oxidation of protein thiols, was to examine whether the oxidation of glutathione depends on Ero1p activity. The rate of glutathione oxidation in yeast cells was assayed by treating cells with the membrane-permeant reductant dithiothreitol to lower their GSSG content, transferring the cells to a fresh medium without dithiothreitol, and then determining the (GSSG) : (GSH) ratio over time as GSSG levels were regenerated by the cell [93].

Using a set of isogenic strains with different levels of Ero1p activity, the authors observed that the rate of GSSG regeneration correlated with that activity [93]. These results show that, in addition to its role in protein-thiol oxidation, Ero1p is the main source of oxidizing equivalents for glutathione in the endoplasmic reticulum. In the Δ*gsh1* strain, the oxidative folding of carboxypeptidase proceeded with normal kinetics but was highly sensitive to oxidative stress, consistent with the role of glutathione as a net reductant [93]. The authors concluded that one possible function of such reducing activity could be to buffer the endoplasmic reticulum against hyperoxidizing conditions by consuming excess oxidizing equivalents through the conversion of GSH to GSSH.

Which role does glutathione normally play during oxidative protein folding *in vivo*? Studies of oxidative protein refolding *in vitro* have shown that both oxidizing and reducing equivalents are necessary for efficient refolding [65]. PDI is likely to catalyze the dithiol-disulfide exchange between GSH and protein disulfides, as it accelerates such reactions *in vitro*. The net effect of an exchange of this kind would be to return some of the disulfide bonds generated by the Ero1p-dependent oxidation pathway to the dithiol state [93]. Given that the isomerization of non-native disulfide bonds is one of the important functions of PDI, a question arises as to how the oxidase and isomerase activities of this enzyme may be integrated in order to expedite the formation of native protein disulfide bonds. In recent years, experimental support for the notion that reduced glutathione may play a direct role in the reduction of disulfides within the endoplasmic reticulum has been obtained [95-98].

1.6.2. GSH Imported into the Endoplasmic Reticulum from Cytosol Can Directly Reduce PDI

Apart from PDI, other oxidoreductases such as EPp57 (a mammalian PDI-like protein [99]) are thought to act as reductases or isomerases and must therefore be maintained in a reduced state to remain active. One candidate for the reduction of the oxidoreductases is GSH. However, as noted above, the glutathione buffer in the endoplasmic reticulum is found at (GSH) : (GSSG) ratios ranging between 1:1 and 3:1, far more oxidizing than in the cytosol; where the ratio is thought to range from 30:1 to 100:1 [19]. Consequently, it has often been speculated that some kind of an intercompartmental compensatory pathway should exist in eukaryotic cells to allow access of cytosolic GSH to the endoplasmic reticulum. To address this question, Jessop and Bulleid [97] examined the redox state of a number of oxidoreductases *in vivo* by modifying free cysteines with an alkylating agent. It was shown that *in vivo* most oxidoreductases exist in a reduced form, suggesting that they might act as isomerases or reductases.

Following oxidative stress, the oxidoreductases are quickly reduced, which proves the existence of a robust reductive pathway in mammalian cells. Using EPp57 as a model, the authors showed that the reductive pathway is cytosol-dependent, and the component responsible for the reduction is GSH [97]. Notably, the recovery of reduced EPp57 was abolished when an inhibitor of glutathione reductase was added to the cytosol, demonstrating that the reduction of GSSG to GSH is required to reconstitute the reductive pathway. In addition, GSH was found to form a mixed disulfide with ERp57 in intact cells [100]. Collectively, the results provided evidence for cytosol being the source of reduced glutathione, which is imported into the endoplasmic reticulum and directly reduces the oxidoreductases.

This view is supported by recent data from Chakravarthi et al. who showed that even though a lowering the level of GSH in the cell leads to accelerated folding, it also leads to the formation of non-native disulfide bonds, which is why GSH may be required for isomerization [95]. Consistent with a role of glutathione as a net source of reducing equivalents is the finding that GSH, rather than GSSG, is selectively imported into endoplasmic reticulum microsomes [101]. The evidence of a robust pathway for the reduction of the oxidoreductases highlights the importance of reduction and isomerization pathways in the cell [97]. Therefore, the redox conditions within the endoplasmic reticulum lumen serve to maintain a balance between reducing equivalents arriving from the cytosol as GSH or as newly synthesized protein thiols, and between the oxidation of protein thiols by Ero1p [96] or by any other potential oxidase [95]. Segregation of the oxidative and reductive pathways is thus achieved owing to the fact that GSH is a poor substrate for Ero1p [77] and that Ero1p can oxidize proteins even at 2.5 mM GSH [103, 97].

The functional analysis of Ero1p, oxidoreductases and glutathione offered a new perspective on how protein disulfide bond formation occurs in the endoplasmic reticulum [75]. An important implication of the view that disulfide bond formation proceeds by sequential transfer of oxidizing equivalents between proteins rather than by transfer from GSSG, is that the flow of oxidizing equivalents might be controlled to a greater extent by the kinetics of protein-protein interactions than by equilibration of protein dithiols and disulfides with the glutathione redox buffer. Therefore the actual redox status of a protein in the endoplasmic reticulum might be determined primarily by its relative reactivity with other redox-active proteins and might differ significantly from that predicted from equilibrium measurements of redox potential relative to glutathione [75].

Chapter 2

Disulfide Bond Formation and Isomerization in Prokaryotes

The periplasm of bacteria is similar in many ways to the endoplasmic reticulum of eukaryotes. In both systems, proteins are synthesized in the cytosol and are translocated in an unfolded form across a lipid bilayer [67, 104]. Protein folding in the periplasm is often accompanied by the formation of disulfide bonds [105]. Yet physical differences between the endoplasmic reticulum and the bacterial periplasm do exist. As opposed to endoplasmic reticulum, the periplasm of Gram-negative bacteria is separated from the extracellular environment by a porous membrane which allows passive diffusion of small molecules such as glutathione. As a consequence of its exposure to the external environment and the lack of an effective redox buffer, bacterial periplasm is prone to much greater variations in its oxidative conditions and pH than the endoplasmic reticulum [105]. It was perhaps in order to overcome these constraints that two separate pathways evolved in the bacterial periplasm: one for disulfide bond formation and another one for disulfide bond isomerization. In the latter case, reducing equivalents necessary for isomerase activity are transferred from the cytoplasm via a complex pathway involving three enzymes (see below).

2.1. A Pathway for Disulfide Bond Formation in the Periplasm

2.1.1. DsbA, a Catalyst in Disulfide Bond Formation

The Discovery of DsbA and Its Biological Significance

The *E. coli* periplasm contains a family of disulfide bond-forming (Dsb) proteins that catalyze disulfide bond formation and rearrangement. The first protein shown to be involved in disulfide bond formation was DsbA. This enzyme was simultaneously identified by two separate groups [14, 106, 107] employing different genetic approaches. Bardwell et al. described the non-essential *dsbA* gene as coding for a 21 kDa periplasmic protein with a C30-P31-H32-C33 motif characteristic of CXXC active sites of the thiol-disulfide oxidoreductase family [14]. Although *dsbA* is not essential for cell viability, DsbA-catalyzed disulfide bond formation is crucial for the correct folding and stability of many periplasmic proteins [14, 107-110]. It has also been reported that *DsbA* mutants cannot fold halocytochrome c [111] or the FlgI component of the flagellar motor, which renders *dsbA*⁻ strains unable to swim [112]. A vivid example of the role played by DsbA is its involvement in the proper folding of periplasmic chaperone PapD. This immunoglobulin-like pilus chaperone features an unusual intrasheet disulfide bond between the last two β-strands of its carboxy-terminal domain. The DsbA-dependent formation of this disulfide bond was shown to be critical for PapD's proper folding *in vivo* [113] and thence for the fulfillment of the cellular functions of this chaperone (see Chapter 5 for details). Accordingly, adhesive P pili of uropathogenic *E. coli* were not assembled by a strain that lacked periplasmic disulfide isomerase DsbA [113].

DsbA Is the Strongest Protein Disulfide Oxidant Known

DsbA is the most oxidizing of all known proteins. It functions primarily by introducing disulfide bonds directly into a protein molecule [114, 115], but is much less effective than PDI at catalyzing intramolecular disulfide rearrangement [116]. DsbA is a 21 kDa monomeric protein which possesses a Cys-Xaa-Xaa-Cys motif typical for thioredoxin-like oxidoreductases. Like the other members of this family, the enzyme has two redox states, a reduced state and an oxidized one. Biochemical studies have identified two features that may predominantly account for the mechanism of DsbA. First, the extremely low pK_a of the thiol group of the N-terminal Cys30 within the Cys-Pro-His-Cys motif makes it highly nucleophilic and prone to leaving its disulfide [117], and second, the reduced enzyme is more resistant to denaturation than both the oxidized one [114, 115] and the mixed disulfide intermediate with thiol-containing substrates [114, 117].

The difference in stability of the two DsbA redox states described by Wunderlich et al. [115] has been suggested to arise from an energetically unfavorable conformation of the oxidized form. The authors compared the guanidinium chloride-induced folding/unfolding equilibrium of the reduced and the oxidized form of the enzyme. The transitions at pH 7.0 and 30°C were found to be fully reversible and allowed a calculation of the free energy of stabilization of oxidized and reduced DsbA according to a two-state model for the unfolding transition. The analysis revealed that reduced DsbA is more stable than oxidized DsbA by 22.7 kJ/mol. The conformational tension of 22.7 kJ/mol in oxidized DsbA quantitatively explained the oxidizing properties of the protein. This led Wunderlich et al. to conclude that the oxidizing properties of DsbA result mainly from the tense conformation of its oxidized form, which is converted to a relaxed, reduced state upon oxidation of thiols by DsbA [115].

In a parallel study by the Creighton group, evidence was obtained for the notion that the disulfide bond of DsbA is so unstable in the folded state that its stability increases by several kcal/mol when the protein unfolds [114]. To explain it, the authors suggested that the disulfide bond destabilizes the folded conformation of DsbA, owing to intramolecular interactions which stabilize the thiol groups in the reduced form and create an unfavorable environment for any disulfide involving Cys30. In the native DsbA conformation, the two cysteine residues possess strikingly different chemical properties, despite their similarity in the unfolded protein and in short peptides [117, 118]. The N-terminal cysteine of the active site, Cys30, is reactive toward alkylating reagents and has a phenomenally low pK_a of just under 3, compared with normal pK_a values of about 9 for cysteine residues [117]. The pK_a of Cys33, in contrast, is abnormally high (10) [117]. Owing to its low pK_a value, Cys30 exists almost solely in the thiolate anion state across the entire physiological pH range and is therefore very reactive. This drives the reaction towards the reduction of DsbA and the oxidation of the unfolded protein [107]. The very low pK_a value of Cys30 also allows DsbA to be involved in disulfide exchange reactions at acidic pH, the reaction rates being not very different from those at neutral pH. This allows disulfide bond formation to occur even when the bacterium is living in an acidic medium [119]. Why is the active site Cys30 of DsbA so reactive? To clarify this problem, three-dimensional structure analysis and site-directed mutagenesis studies proved helpful.

DsbA Structure. The Origin of Its Exceptional Oxidizing Power

The crystal structure of oxidized DsbA from *E. coli* was first determined at 2 Å resolution [120] and later refined to 1.7 Å resolution [121]. Despite very low sequence similarity with thioredoxin, a thioredoxin-like domain was found. An additionally inserted 76-residue domain, composed essentially of α-helices, covers part of the active site. While the conformations of the residues forming the active sites of oxidized DsbA and thioredoxin are the same, the presence of this extra domain results in a different environment for the N-terminal active site Cys in the two proteins. The structures of both oxidized and reduced DsbA forms were determined, and their comparison revealed local conformational changes occurring in the active site in response to disulfide reduction, giving rise to a network of stabilizing interactions around the active site thiolate [121].

To shed more light on the nature of these interactions, Graushopf et al. sought to answer the question of whether the two central residues in the Cys-Pro-His-Cys motif are able to alter the redox potential of DsbA by affecting the pK_a of Cys30. Site-directed mutagenesis was employed to exchange randomly the two residues separating the active site cysteines, followed by an analysis of the resulting mutants' oxidizing properties [122]. To compare the oxidizing power of different proteins, the equilibrium constant (K_{ox}) for forming disulfides by exchange with glutathione redox buffer is commonly used [123]. The K_{ox} value for the active site disulfide of DsbA is about 0.1 mM, making it a very potent donor of disulfide bonds. For comparison, thioredoxin, which is thought to function as a dithiol reductant in the cytoplasm, has a K_{ox} value of about 2 M at pH 7.5 [29].

Determination of the relative oxidizing power of nine mutants showed that redox equilibrium constants span a wide range from 0.1 mM for wild-type DsbA to about 200 mM for a mutant with a Cys-Pro-Pro-Cys sequence. Such range of redox potentials indicates that the two central residues are critical in determining the redox properties of DsbA. The most reducing mutant obtained had the sequence of *Arabidopsis* thioredoxin [124] and a redox potential approaching that of *E. coli* thioredoxin, which had been measured at about 2 M [29]. This shows that one can change DsbA from a very oxidizing protein to one with a thioredoxin-like potential simply by altering the active site sequence to mirror that of thioredoxin [122]. The pK_a of Cys30 was found to be dramatically affected in the mutants obtained. The most reducing mutant, Cys-Pro-Pro-Cys, had a measured pK_a of approximately 6.7, over 3 pH units higher than that of the wild type. Notably, the pK_a was found to vary in proportion to K_{ox}, suggesting that pK_a plays a very important role in determining K_{ox}. In this way, evidence was obtained that the two central residues within the active site play a critical role in determining both the oxidative power of DsbA and the pK_a of Cys30.

2.1.2. DsbB, a Protein Responsible for Reoxidation of DsbA

After transferring its disulfide bond to the target protein, DsbA needs to be reoxidized to continue serving as a source of potential for forming disulfide bonds. The first signs pointing to the existence of a protein which is responsible for DsbA reoxidation came from the observation that in the periplasm of *dsbB* mutant bacteria, DsbA accumulates mostly in the reduced form whereas in wild type bacteria both reduced and oxidized species are present [125]. These results were consistent with independent studies [109, 112] reporting the identification of the *dsbB* gene. The fact that mutations in *dsbA* and *dsbB* produce similar defects in disulfide bond formation suggests that both proteins are on the same major oxidation pathway. A model predicting that DsbB could reoxidize reduced DsbA protein, was then proposed [125].

How Does DsbB Reoxidize DsbA?

DsbB is a 21 kDa inner-membrane protein that was predicted to have four transmembrane helices and two functionally important periplasmic loops [126-129]. In addition, DsbB has been shown to bind quinone with at least one high-affinity binding site [130, 131]. Predicted DsbB structure is represented on Figure 7.

Each periplasmic region contains a pair of essential cysteines, Cys41 and Cys44 in the N-terminal periplasmic loop (P1) and Cys104 and Cys130 in the C-terminal periplasmic domain (P2) [134]. The first pair of cysteines is located in a C-X-X-C arrangement, which is reminiscent of the C-X-X-C motif present in many thioredoxin-like proteins. Recently, the crystal structure of the DsbB-DsbA complex was determined at 3.7 Å resolution [135].

In accordance with the earlier prediction, the structure of DsbB revealed four transmembrane helices with both termini oriented toward the cytoplasm. A short helix (residues 112-126) with a horizontal axis also exists in the periplasmic region of DsbB, forming part of the long loop that connects helices 3 and 4 [135]. Whereas DsbB in the resting state contains a Cys104-Cys130 disulfide, Cys104 in the binary DsbB-DsbA complex appears engaged in the intramolecular disulfide bond and captured by the hydrophobic groove of DsbA, resulting in separation from Cys130. This cysteine relocation allows Cys130 to approach and activate the disulfide-generating reaction center composed of Cys41, Cys 44, Arg 48, and ubiquinone. In sum, the structural data provided evidence for a rearrangement of the periplasmic regions of DsbB induced by DsbA, and contributed significantly to the understanding of the mechanism whereby DsbA is kept oxidized in spite of its strong oxidizing (reduction-prone) property.

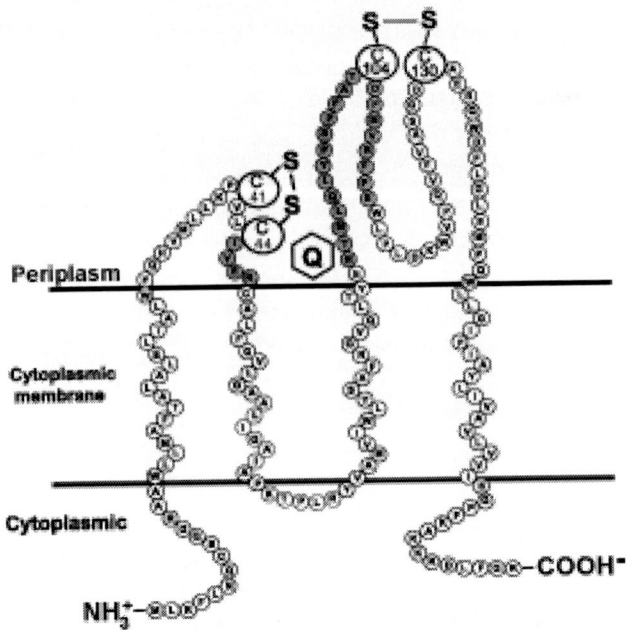

Figure 7. Predicted membrane topology of DsbB [132,133]. Residues that are believed to be involved in quinone binding are from both periplasmic loops and are shaded grey. Q stands for quinone.

A proposed pathway of DsbA oxidation is represented in Figure 8. The Cys104-Cys130 disulfide bond in DsbB's second periplasmic domain is thought to be the disulfide donor to DsbA. It is believed that reduced DsbA forms a complex with DsbB, in which Cys30 of DsbA and Cys104 of DsbB become disulfide-bonded. This mixed disulfide [136] is probably an intermediate in the process of DsbA reoxidation (Figure 8, step 2). When Cys33 of DsbA is replaced by a serine, a mixed disulfide between DsbA and DsbB accumulates because this complex cannot be resolved by Cys33 [129].

After the resolution of the mixed disulfide bond by an attack of DsbA's Cys33, DsbA is released in oxidized form. At the same time, DsbB's reduced Cys130 attacks Cys41-Cys44 to form the Cys130-Cys41 disulfide, thus rendering Cys44 reduced. Its thiolate form will then interact with ubiquinone [128]. In agreement with this model, a systematic examination of the redox states of DsbB's cysteine residues revealed that while DsbB has two disulfide bonds under normal aerobic conditions, oxidation of Cys104 and Cys130 requires the presence of an N-terminally located Cys41-Cys44 disulfide bond [137]. This provides the first link between disulfide bond formation and the electron transport chain.

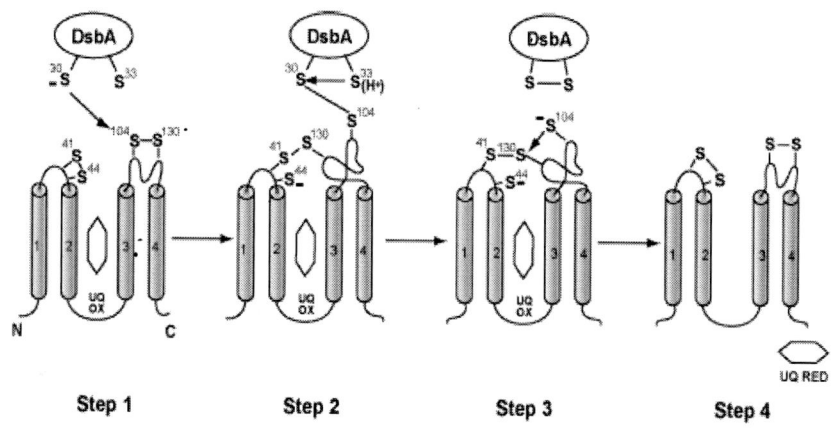

Figure 8. Proposed pathway of DsbB-catalyzed DsbA oxidation [140]. Electron transfer from reduced DsbA to DsbB and from there to ubiquinone (UQ) can be divided into several steps. In step 1; reduced Cys30 of DsbA attacks the Cys104-Cys130 disulfide of DsbB to form a Cys30-Cys104 intermolecular disulfide. In step 2, the resulting reduced Cys130 attacks Cys41-Cys44 to form a Cys130-Cys41 disulfide, thereby reducing Cys44, whose thiolate form will then interact with ubiquinone to induce the red-shifted electronic state. In step 3, the DsbB-DsbA complex is resolved, releasing DsbB with its Cys 44 and Cys 104 reduced. It is not known how the less reactive Cys33 residue of DsbA [44] becomes activated to initiate the resolution. Finally (step 4), a regeneration of fully oxidized DsbB takes place, coupled with the reception of two electrons by the Cys44-activated ubiquinone, which is now reduced to ubiquinol and later subjected to oxidation by terminal oxidases. See text for details. UQ OX, ubiquinone; UQ RED, ubiquinol.

An *in vitro* reconstitution of the disulfide bond catalysis system using purified components verified the connection between electron transport and disulfide bond formation and showed that ubiquinone serves as an immediate electron acceptor of DsbB [127]. One of the most significant results of that study was the finding that DsbB uses components of the electron transport system to drive its reoxidation: DsbB transfers its electrons on to oxidized ubiquinone, which then donates them to cytochrome oxidases, which reduce oxygen. Therefore, DsbB uses the oxidizing power of quinones to generate disulfides *de novo*. This unique catalytic activity can be regarded as the major source of disulfides in prokaryotes [133, 138-140]. In this reaction, DsbB has two substrates: DsbA and quinones, and can be described as an enzyme catalyzing a two-substrate reaction:

$$\text{DsbA}_{\text{dithiol form}} + \text{ubiquinone} \xrightarrow{\text{DsbB}} \text{DsbA}_{\text{disulfide form}} + \text{ubiquinol}$$

In the absence of DsbB, almost no oxidation of DsbA occurred; oxidation was not accelerated by the addition of either ubiquinone or DsbB alone. These controls show that ubiquinone cannot directly reoxidize DsbA [127]. Detailed examination of the ubiquinone reductase activity of DsbB revealed that DsbB is capable of accelerating the redox reaction specifically and very efficiently. Thus, the apparent k_{cat}/K_m value for the DsbB-catalyzed reaction between ubiquinone and reduced DsbA is $4 \cdot 10^6$ $M^{-1}s^{-1}$ [140].

A Disulfide Bond across the N-Terminal Pair of Cysteines (Cys41-Cys44) in DsbB Is Formed by Ubiquinone Reduction

As considered above, it is very likely that ubiquinone directly reoxidizes the Cys41-Cys44 pair of cysteines [140]. In support of this concept, DsbB was found to be capable of inducing an electronic transition in the bound ubiquinone molecule [141]; this transition was characterized by a spectacular emergence of an absorbance peak at 500 nm, giving rise to a visible pink color. The ubiquinone red shift was observed by stopped-flow rapid scanning spectroscopy during the reaction between DsbA and DsbB. Mutation and reconstitution experiments established that the unpaired Cys at position 44 of DsbB is primarily responsible for the chromogenic transition of ubiquinone, and this property correlates with the functional arrangements of amino acid residues in the neighborhood of Cys44. The observed pH-dependence in the color development is consistent with the deprotonated thiolate anion at position 44 interacting non-covalently with a specific region of ubiquinone to modulate its electronic states. The authors propose that the Cys44-induced anomaly in ubiquinone represents its activated state, which drives the DsbB-mediated electron transfer [141].

Taken together, the available information is consistent with the DsbA oxidation pathway shown schematically in Figure 8. The process probably proceeds through the disulfide-linked DsbA-DsbB intermediate, which is resolved as the rate-limiting stage, to yield oxidized forms of DsbA and DsbB while reducing ubiquinone. This reaction is initiated by the formation of an intermolecular Cys104-Cys30 bond between DsbB and DsbA, resulting in the liberation of reduced Cys130, which in turn triggers a disulfide rearrangement within DsbB to form a Cys41-Cys130 interloop disulfide [142], and a consequent reduction of Cys44. Furthermore, the reduced Cys44 induces an electronic transition of bound ubiquinone, leading to the resolution of the Cys104-Cys30 bond, regeneration of all the disulfide bonds in the system and reduction of ubiquinone. The completion of the cycle is coupled with transition and electron acceptance by ubiquinone [141, 143].

2.2. Similarities in Prokaryotic and Eukaryotic Disulfide Bond-Forming Pathways

Figure 9 summarizes the results obtained by Bader et al. in their study of the oxidative protein folding system, which showed how disulfide bond formation is linked to metabolism in a bacterial cell [127]. In particular, reoxidation of DsbB was shown to be dependent on the presence of either bd or bo cytochrome oxidases (functionally similar to eukaryotic cytochrome oxidase), and of either a menaquinone or a ubiquinone electron acceptor.

Electrons flow via cytochrome bo oxidase to oxygen under aerobic conditions or via cytochrome bd oxidase under partially anaerobic conditions. Under truly anaerobic conditions, menaquinone shuttles electrons to alternate final electron acceptors such as fumarate. These findings provided a satisfactory explanation for the ability of the Dsb system to function efficiently in promoting disulfide bond formation under anaerobic growth conditions [144]. What can be the source of oxidation potential when oxygen is not present? Now that menaquinone has been identified as an effective recipient of electrons from DsbB, it becomes clear that DsbB can use its primary electron acceptors depending on the degree of aerobiosis [127, 145].

The pathway for protein disulfide bond formation in the bacterial periplasm provides a useful analogy to the protein oxidation system in the eukaryotic endoplasmic reticulum considered earlier in this chapter. Indeed, the pathways for disulfide bond formation in the lumen of the endoplasmic reticulum of eukaryotic cells and in the periplasmic space of prokaryotic cells are similar in their general outlines. In both cases, disulfide bonds are introduced into folding proteins by transfer from a disulfide bond carrier that is a member of the thioredoxin family of proteins. In eukaryotic cells, the major disulfide bond carrier is the soluble protein PDI, whereas in bacteria it is another soluble protein, DsbA. Mutations in both *PDI* and *dsbA* disrupt the oxidation of secretory proteins, and PDI can complement mutations in *dsbA* when targeted to the bacterial periplasm [73, 14, 146].

The disulfide bond carrier proteins, in turn, receive disulfide bonds from another class of protein thiol-oxidoreductases. In both eukaryotic and prokaryotic systems, this function is fulfilled by membrane-associated oxidoreductases (flavoenzyme Ero1p in the former case and DsbB in the latter case). As noted by Frand et al. [20], the similarities between the key components of eukaryotic and prokaryotic disulfide bond-forming systems are striking. A different picture is observed in the case of disulfide bond isomerization pathways, which appear to be quite different in prokaryotes and eukaryotes.

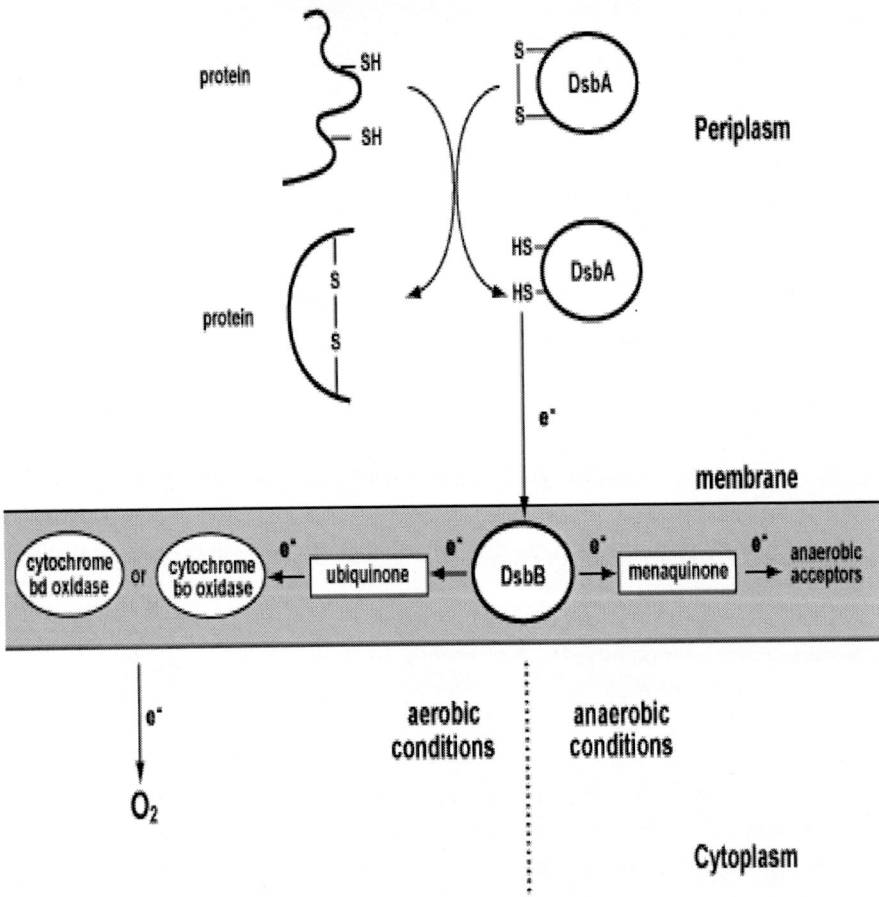

Figure 9. The pathway of disulfide bond formation in *E. coli* periplasm. Electrons (e^-) flow from newly synthesized peptides in the periplasm to DsbA upon disulfide bond formation. Reduced DsbA is reoxidized by membrane-bound DsbB. Under aerobic conditions, DsbB passes electrons to ubiquinone, and these electrons are ultimately transferred to oxygen by cytochrome oxidases. Under anaerobic conditions, DsbB passes its electrons to menaquinone, which is then reoxidized by anaerobic oxidases. The end result is a reoxidized DsbB that is ready for another round of reoxidizing DsbA. Adapted from Bader et al. [127] and from Tan and Bardwell [105].

2.3. A Pathway for Disulfide Bond Isomerization in the Periplasm

To a protein with multiple disulfides, the DsbA-DsbB system does not always supply the correct disulfides. Thus, in an *in vitro* refolding experiment on RNase A, Bader et al. [141] demonstrated that, although DsbA can catalyze a complete oxidation of RNase A, this does not result in the appearance of catalytically active enzyme species. The inactivity of refolded RNase A is most probably caused by an introduction by DsbA of non-native disulfides, resulting in a misfolding of the substrate protein [141, 147]. When this experiment was performed in the presence of glutathione redox buffer, refolded RNase A proved catalytically active, suggesting the presence of a reducing agent is a prerequisite for productive refolding. Since glutathione is not found in *E. coli* periplasm and is thus a non-physiological reoxidant, a question arises as to mechanisms that allow proteins with multiple disulfides to fold correctly *in vivo*. The first step towards answering this question was with the discovery in 1994 of an enzyme designated DsbC. Two research groups independently isolated the *dsbC* gene using different genetic approaches [148, 149]. DsbC was identified as a periplasmic protein displaying thiol-disulfide oxidoreductase activity both *in vivo* and *in vitro* [148, 149].

2.3.1. DsbC, a Disulfide Isomerase

Mature DsbC is a stable dimer of identical 23 kDa subunits, which appear to function independently. Each subunit has a Cys98-Gly99-Tyr100-Cys101 segment that forms an unstable and reactive disulfide bond; only the first cysteine residue is accessible, much the same as in thioredoxin and DsbA. The nucleophilic character of Cys98 is the result of this residue's extremely low pK_a value, equal to 4.1 [63]. The other two cysteine residues in DsbC form a buried, structural disulfide bond [150]. A comparative study of the reactivities and stabilities of the active site disulfide bond in DsbC and DsbA showed that both are very unstable and can be transferred rapidly to reduced proteins and peptides. A number of observations (DsbC is much more active than DsbA in catalyzing protein disulfide rearrangements; inactivation of the *dsbC* gene inhibits disulfide formation mainly in proteins with multiple disulfide bonds, which are most likely to be dependent upon disulfide rearrangements) lent credibility to the suggestion that the disulfide isomerase activity of DsbC may be its principal function *in vivo* (Zapun et al., 1995 [150]). A year later, the first direct evidence for an *in vivo* role of DsbC in disulfide bond isomerization was presented. According to the results obtained by Rietsch et al., DsbC is not involved in *de novo* disulfide bond formation in the cell [151]. These data, together with results from other laboratories [152, 153], supported the notion that DsbA is the main oxidase in the periplasm, while DsbC is required for the isomerization of disulfide bonds.

For DsbC to attack incorrectly formed disulfide bonds in substrate proteins, its two active site cysteines must be in the reduced form. In a further study, Rietsch et al. demonstrated that this is in fact the case: in wild type cells, DsbC is present in a reduced state [154]. The proposed mechanism of DsbC action can be summarized as follows. Cys98 of DsbC forms a mixed disulfide with an incorrectly paired cysteine in a protein. This mixed disulfide is then resolved either by an attack from another cysteine in the misfolded protein, leading to the formation of a more stable disulfide in the substrate and a reduced DsbC, or by an attack from Cys101 in DsbC. In the latter case, DsbC becomes oxidized and needs reduction in order to be recycled [105, 107].

Using simple unstructured peptides with one or two cysteine residues, Darby et al. have shown that reactions between the various forms of peptide and DsbC proceed up to 10^6 times faster than those involving glutathione and DsbC, and mixed disulfide complexes of DsbC and a peptide are 10^4 times more stable than the corresponding mixed disulfides with glutathione. These observations suggest that non-covalent binding interactions between the peptide and DsbC contribute to the very rapid kinetics of substrate utilization. Interactions between DsbC and a peptide were found to be more stable than those between DsbA and the same peptide, which helps explain the superior isomerase activity of DsbC [155].

To investigate the catalytic properties of DsbC and DsbA during the folding of proteins with very complex disulfide bond patterns, Maskos et al. [156] selected the bifunctional α-amylase/trypsin inhibitor (RBI) from *Ragi* as a model substrate. RBI is a 13.1 kDa protein with five overlapping disulfide bonds. Under oxidizing redox conditions, RBI folding is hampered by the accumulation of a large number of intermediates with non-native disulfide bonds, while a surprisingly low number of intermediates accumulates under optimal or reducing redox conditions. DsbC catalyzes the folding of RBI under any redox conditions *in vitro*, but is particularly efficient in rearranging buried, non-native disulfide bonds formed under oxidizing conditions. In contrast, the influence of DsbA on the refolding reaction is essentially restricted to reducing redox conditions, where disulfide formation is rate-limiting. The effects of DsbA and DsbC on the folding of RBI in *E. coli* are very similar to those observed *in vitro*. This indicates that DsbC is the bacterial enzyme of choice for improving the periplasmic folding yields of proteins with very complex disulfide bond patterns [156].

2.3.2. Dimerization of DsbC Is a Prerequisite for Its Isomerase Activity

Determination of the three-dimensional structure of DsbC showed it to be a V-shaped protein, with each monomer forming one arm of the V [62]. Each monomer is subdivided into two domains: an N-terminal dimerization domain and a C-terminal catalytic domain with a thioredoxin-like fold, joined via hinged linker helices. The N-terminal domain from each monomer forms the dimer interface at the base of the V through β-sheet hydrogen bonds. The active sites from each monomer face each other in the interior of the V. The surface of this V-shaped cleft is composed mainly of hydrophobic and uncharged residues, suggesting that it may be involved in the binding of substrate proteins [62, 107]. An extended surface for peptide binding is a feature important for DsbC's isomerase activity, which involves conformational rearrangements of the substrate protein. DsbC has been shown to exhibit chaperone activity that promotes reactivation and suppresses the aggregation of denatured D-glyceraldehyde-3-phosphate dehydrogenase [157].

Limited proteolysis experiments confirmed the role of the N-terminal domain in the association of individual DsbC monomers into a dimer [63]. Removal of the N-terminal domain produced a compact and stable C-terminal fragment which retained the active site sequence, had a pK_a of active site thiols very close to that of DsbC, and showed only minor differences in conformation compared to an intact DsbC molecule. However, this fragment was inactive as an isomerase in catalyzing the formation of correct disulfide bonds in scrambled RNase A and in denatured and reduced bovine pancreatic trypsin inhibitor. In contrast to native DsbC, the above fragment exists as a monomer and does not have the chaperone properties normally exhibited by wild type DsbC [63]. These results indicated that dimerization is necessary for isomerase activity, and provided support for the idea that such activity might require the presence of both active sites of the DsbC molecule.

To further explore the possibility that protein disulfide isomerase activity can only be exhibited by a dimeric form of the enzyme, containing two functionally competent active sites, Wang and his collaborators expressed and purified four hybrid dimers composed of different combinations of the following thioredoxin-like domains: DsbA, N-terminal active-site domain a of PDI, non-active site domain b of PDI, C-terminal domain of DsbC, and thioredoxin proper [158]. In these hybrids, the N-terminus of thioredoxin, DsbA, and the a and the b domain of PDI were separately linked to the C-terminus of the association domain by the flexible linker of DsbC. The authors posed the questions of whether the hybrids could form a dimeric structure, and if they could, whether there was any gain of function due to intersubunit interactions within the dimer. The results of this elegant study revealed that hybridization had indeed led to a homodimeric structure for the four hybrids. Moreover, only the hybrids that contained two functioning active sites (i.e. included thioredoxin, DsbA and the a domain of PDI, but not the b domain of PDI) were endowed with isomerase and chaperone activities.

2.3.3. Dimerization of DsbC Protects Its Active Sites from Oxidation by DsbB

The newly revealed architecture of the DsbC molecule shed light on the mechanism behind an interesting phenomenon discovered earlier by Bader et al., who showed the reoxidation of DsbC by DsbB to be at least 500 times slower than the reoxidation of DsbA [140]. It was an important finding because DsbB, by discriminating between DsbC and DsbA, allows for the separation of the oxidative pathway of disulfide bonds formation from the pathway of disulfide bonds isomerization in the periplasm. How does DsbB distinguish between DsbA and DsbC? An examination of the three-dimensional structure of DsbC led the authors to suggest that its active sites, which point toward the interior of the cleft formed by the dimer, are not accessible to DsbB and hence resistant to oxidation by the latter [140].

To prove the molecular origin of the above effect, Bader et al. [159] selected *dsbC* mutants that complemented a *dsbA* null mutant, showing that they could actually replace *dsbA in vivo*. In these mutants, DsbC is present as a monomer, as opposed to dimeric wild-type DsbC. Based on these findings, the authors designed DsbC mutants containing amino acid replacement in the dimerization domain. The monomeric DsbC G49R mutant, as well as a DsbC variant consisting of just the thioredoxin domain, were both able to complement a *dsbA* null for motility and alkaline phosphatase activity *in vivo*. Such complementation is dependent on the presence of DsbB, with monomeric DsbC acting as a substrate for oxidation by DsbB [159]. Thus, monomerization of DsbC was shown to turn a disulfide isomerase into an oxidase, which becomes a substrate for DsbB. These results illustrated how the delicate balance between the oxidative and reductive pathways is controlled by the dimerization of DsbC. They made clear how DsbA and DsbC can exhibit quite different redox activities within the same cellular compartment without interfering with each other [159].

In a recent study by another research group [160], the molecular mechanism of the DsbC's avoidance of misoxidation by DsbB was investigated in more detail. The authors were interested in clarifying the role played in the above effect by the α-helical linker that connects the N-terminal dimerization domain with the C-terminal thioredoxin domain of DsbC. It was shown that DsbC's resistance to oxidation by DsbB is abolished by deletions of one or more amino acids from the α-helix connecting the N-terminal and the C-terminal domains. As a result, mutant DsbC with α-helix deletions could catalyze disulfide bond formation and complement the phenotypes of *dsbA* cells. Meanwhile, the dimeric wild type form of the enzyme was strongly resistant to oxidation by DsbB.

Given that the dimerization region and the thioredoxin catalytic domain of DsbC are able to fold independently and are stable on their own [63], it is unlikely that a truncation in the linker should bring about any major conformational changes in either of these domains. In the authors' opinion, deletion of an amino acid from the linker could be expected either to slightly shorten the α-helix or to cause a rotation of the end of the helix and possibly of the entire catalytic domain, including the active site. This would shorten the distance between the peptide binding cleft and the catalytic center and possibly influence the interaction of the protein with DsbB [160]. Taken together, the results of the above study indicate that it is not the dimeric nature of the enzyme *per se* that keeps the DsbC active site from interacting with the catalytic center of DsbB, but rather some conformational features of a more subtle nature, probably having to do with the geometry and orientation of the active sites in the overall structure of the molecule [160].

2.3.4. DsbG, a Paralogue of DsbC

Another Dsb protein, DsbG, was identified by Andersen et al. [161], who found that the *dsbG* gene, when present in multiple copies, could make DsbB mutants resistant to elevated concentrations of dithiothreitol. DsbG is a 25.7 kDa protein forming a stable periplasmic dimer and showing a 28% sequence identity and a 56% sequence similarity to DsbC [161]. The Cys-Pro-Tyr-Cys sequence in DsbG forms an unstable disulfide bond that is readily reduced by glutathione. Despite the highly oxidizing nature of the periplasmic space, in wild type cells DsbG is found exclusively in the reduced state, as would be expected if DsbG were acting as a disulfide isomerase or reductase [162]. Shao et al. have shown that just like DsbC, DsbG functions as a chaperone in the correct folding of two classical chaperone substrate proteins, citrate synthase and luciferase [163].

Despite this similarity, there are indications that the substrate range of DsbG may be narrower than that of DsbC [164]. Thus, a systematic investigation of the *in vivo* substrates of *E. coli* periplasmic disulfide oxidoreductases DsbA, DsbC and DsbG, did identify two substrates of DsbC: RNase A, a periplasmic ribonuclease, and MepA, a periplasmic murein hydrolase, but did not detect any substrates for DsbG. It was also shown that a *dsbG* null mutation has no effect on the *in vivo* function of RNase A, indicating that DsbC and DsbG do not have identical *in vivo* substrate specificity [165]. A question therefore arises as to why two isomerases, DsbC and DsbG, are encoded in bacteria. To shed more light on the function of DsbG, Heras et al. subjected it to structural studies, which revealed unexpected and surprising features in this enzyme [166].

It was found that DsbG has a significantly longer helical linker than DsbC, which substantially increases the size of the binding site cleft. Conserved acidic residues in DsbG (which are not present in DsbC) form negatively charged patches in the otherwise hydrophobic cleft, which would suggest interaction with globular protein substrates having charged surfaces. Moreover, a longer lip between strands β2 and β3 of DsbG forms a groove incorporating conserved polar/charged residues at the base of the hydrophobic cleft [166]. These differences between DsbG and DsbC indicate that the two proteins interact with very different substrates. The characteristics of the DsbG binding surface are consistent with binding target proteins that are folded or partially folded. This means that the isomerase function of DsbG may be directed at proteins further down the folding pathway than those interacting with DsbC [166].

2.3.5. DsbD, a Recycler of Reduced DsbC and DsbG

The Pathway for DsbC and DsbG Reduction in the Periplasm

In order to be functional as isomerases, DsbC and DsbG need to be kept reduced. *In vivo* data show that both DsbC and DsbG are indeed maintained in reduced form even though they exist in the strongly oxidizing environment of bacterial periplasm [154]. DsbC and DsbG are kept reduced by the action of an inner membrane protein called DsbD (or DipZ). The *dsbD* gene was identified in screens for mutants deficient in cytochrome c synthesis [168], disulfide bond formation in periplasmic proteins [169], and copper resistance [170]. DsbD has a molecular mass of 59 kDa making it the largest protein in the Dsb family [107]. It exhibits a well-defined domain structure, with two periplasmic and one transmembrane domain (Figure 10). Evidence supporting DsbD's essential role in the isomerization pathway comes from genetic work examining the effects of a dsbD null mutant on periplasmic protein folding. In a dsbD null mutant, DsbC and DsbG accumulate in oxidized form [151]. Remarkably, the effect of a double dsbC- dsbD- mutant on alkaline phosphatase expression is not cumulative, which shows that DsbC and DsbD are part of the same pathway [151].

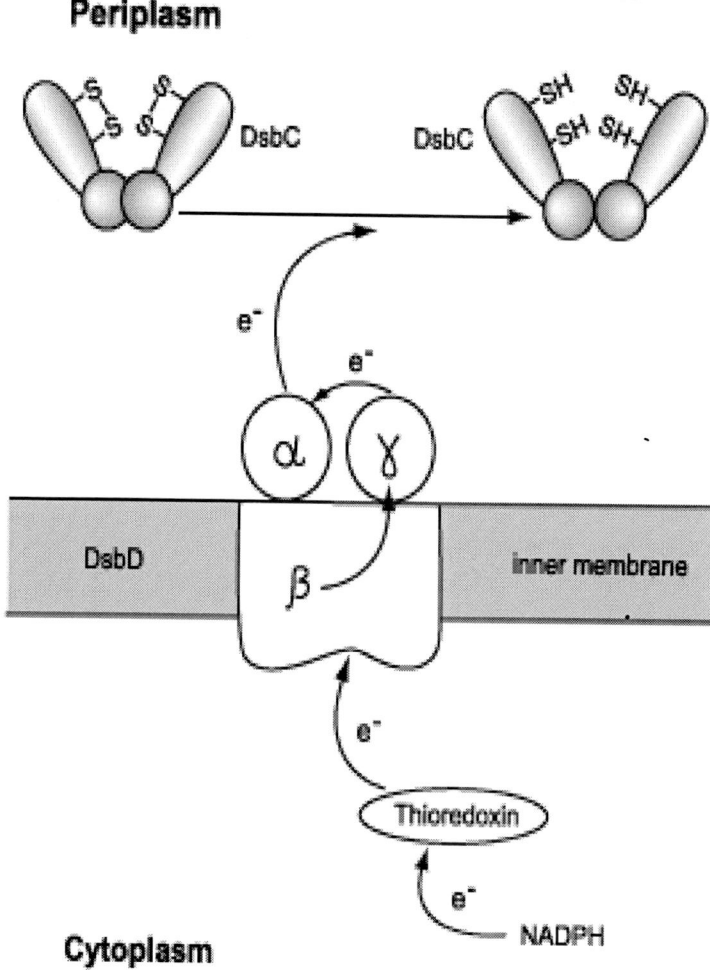

Figure 10. The pathway of DsbC and DsbG reduction in the periplasm. The oxidized form of DsbC accepts electrons which flow from NADPH to thioredoxin reductase, from thioredoxin reductase to thioredoxin, and subsequently to DsbD which is made up of three domains: an N-terminal periplasmic domain (α domain), a hydrophobic transmembrane domain (β domain), and a C-terminal thioredoxin-like domain, also located in the periplasm (γ domain) [167].

In order to reduce DsbC and DsbG, the DsbD protein itself has to be reduced. The proteins required to facilitate DsbD reduction were identified by genetic studies. As a result, the components of the pathway that allows DsbC and DsbG to remain reduced in the periplasm were also identified [151, 154]. The current model for the maintenance of DsbC and DsbG in reduced form involves a passage of electrons from the reducing environment of the cytoplasm to the oxidizing periplasm. In this model (see Figure 10), thioredoxin, which is kept in reduced form by thioredoxin reductase and NADPH, passes electrons to DsbD. DsbD's position in the inner membrane allows it to receive electrons from cytoplasmic thioredoxin and transfer them to periplasmic proteins DsbC and DsbD [147, 107].

Structure of the N-Terminal Periplasmic Domain of DsbD (Termed DsbDα)

The crystal structure of the first (N-terminal) periplasmic domain of DsbD was solved independently in two laboratories [171, 172]. It was found that crystallized α domain assumes an immunoglobulin-like structure, making it the first example of a thiol oxidoreductase with an immunoglobulin fold. A single domain of DsbDα consists of four antiparallel sheets, two of which form a β-sandwich. The β-sandwich consists of two larger antiparallel β-sheets, each of which has three strands. At the N-terminus end of DsbDα, the five-stranded antiparallel β-sheet is reminiscent of a partial β-barrel (β6, β7, β3, β10, β11), which harbors Cys 103 and Cys109, the residues essential for activity [159], see Figure 11. Cys103 and Cys109 form a disulfide bond between β10 and β11 strands. The distance between the two S_γ atoms of Cys103 and Cys109 is about 2Å, consistent with a disulfide bond linking the two residues [173].

A remarkable feature of the active site is the loop region between strands β6 and β7, which lies over the disulfide bond formed between Cys103 and Cys109, completely shielding the active site from the environment. In particular, the aromatic ring of Phe70 within this loop region forms close van der Waals interactions with the S_γ atom of Cys109 and may contribute to the stability of the disulfide bond. The authors suggest that the loop, by capping the active site, protects it from non-specific interactions with other periplasmic components, and that structural displacement of the cap from the active site must precede a reduction of the disulfide bond of DsbDα by the C-terminal domain of DsbD [171, 172].

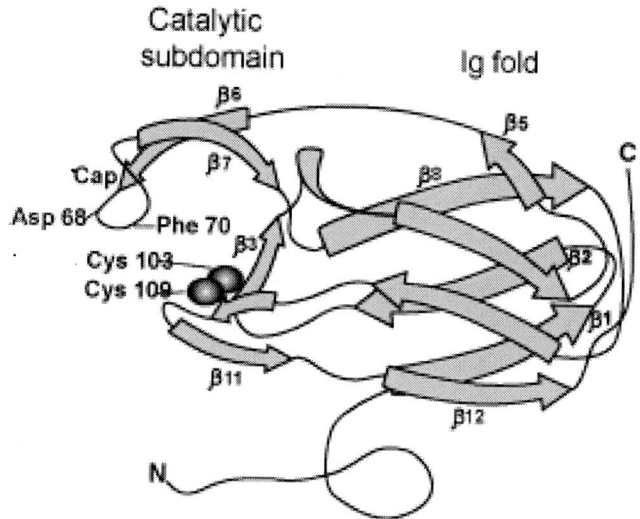

Figure 11. Crystal structure of the DsbD α domain [171, 172]. The overall structure of this domain can be divided in two subdomains: an immunoglobulin (Ig) fold consisting of a β-sandwich formed by two antiparallel β-sheets, and a catalytic subdomain which is inserted into the antigen binding end of the Ig fold. The catalytic subdomain contains the active site cysteine residues Cys103 and Cys109. The catalytic β-sheet is completed on the opposite side of β3 by strands β6 and β7. These strands and the connecting loop between them form a cap structure over the active site; Phe70 located in this cap establishes an aromatic-sulfur interaction with the S_γ atom of Cys109, and may contribute to the stability of the disulfide bond.

Structural Characteristics of the Transmembrane Domain of DsbD (Termed DsbDβ)

The α-domain of DsbD is followed by a hydrophobic core comprising eight predicted transmembrane-spanning α-helices that form the β-domain [168-170]. A detailed analysis of DsbD membrane topology has been performed by Chung et al. [167]. Fusions of DsbDβ membrane-spanning segments to the mature part of alkaline phosphatase (PhoA) resulted in hybrid proteins that either translocated the C-terminal PhoA domain into the periplasm or remained exposed to the cytoplasmic environment. As PhoA requires DsbA-catalyzed disulfide bond formation in the periplasm for enzymatic activity [108], active PhoA fusions point to membrane translocation, whereas inactive fusions indicate that the reporter remained exposed in the bacterial cytosol [167]. 15 fusions were constructed that tethered the PhoA domain to various predicted intramembranous segments of DsbD. The results of this investigation revealed that Cys163 located at the end of transmembrane helix 1 is positioned close to the cytoplasmic side on the lipid bilayer; it thus appears poised to accept electrons from thioredoxin and may somehow direct them towards the periplasmic dithiol motifs of DsbD. Consistent with this function of Cys163 is the observation that this residue is found conserved in all known DsbD homologues [167, 174].

In a parallel study, Gordon et al. provided evidence that besides Cys163, another highly conservative cysteine residue, Cys285, located within membrane-spanning regions of DsbD (in the fourth helix), is also functionally important [175]. This supported the idea that a pair of Cys163-Cys285 residues might play a role in the transfer of electrons across the cytoplasmic membrane [167, 174, 176]. However, the predicted location of the essential Cys163 and Cys285 on the opposite sides of the membrane posed several unresolved questions about the mechanism of transmembrane electron transport. In particular, it remained to be proven that Cys163 and Cys285 do actually form a disulfide bond, and an explanation was needed of how electron transfer can proceed between these residues if they face opposite sides of the membrane.

By using a variety of experimental and theoretical approaches, Katzen and Beckwith managed to shed some light on these problems [177]. They came up with strong evidence that Cys163 and Cys285 are capable of forming an intramolecular disulfide bond, strengthening an earlier suggestion that the electron transfer mechanism is based on a series of thiol - disulfide bond exchange reactions [178]. At the same time, the data obtained in this investigation led the authors to conclude that both essential cysteines are probably solvent-exposed to the cytoplasmic side of the inner membrane. This finding contradicts previous topological models that placed these residues on opposite sides of the membrane and raises a new question regarding the putative mechanism of transmembrane electron transfer. While the above results are consistent with a reaction between thioredoxin and the β-domain of DsbD, the question concerning the transfer of reducing potential from β to γ still has no answer [177].

Structure of the Second Periplasmic Domain of DsbD (Termed DsbDγ)

The crystal structure of DsbDγ was first determined by Kim et al. at 1.9 Å resolution [179]. It was found that each DsbDγ molecule contains six α-helices and four β-strands, resulting in a thioredoxin fold. The active site motif (Cys461-Val462-Ala463-Cys464) is located at the terminus of helix α2a. A disulfide bridge between the two active site cysteines was evident in the electron density map, consistent with the non-reducing condition at the crystallization step [179]. However, some important questions remained to be explored: What is the structure of the reduced form? Are there conformational changes upon the oxidation-reduction of the active site disulfide? What are the pK_a values of the essential cysteine residues? And, finally, what is the thermodynamic stability of the oxidized and reduced forms? In a recent study, Stirnimann et al. [180] determined four high-resolution structures of DsbDγ: oxidized, chemically reduced, photo-reduced, and chemically reduced at pH levels increased from 4.6 to 7. Notably, the latter structure was refined to 0.99 Å resolution, the highest achieved so far for a thioredoxin superfamily member.

Using a combination of high-resolution protein crystallography, biochemical experiments and computational approaches, the authors were able to uncover unprecedented structural details of DsbDγ, demonstrating that the domain is very rigid and hardly undergoes any conformational change upon disulfide reduction or interaction with DsbDα. The very similar thermodynamic stabilities of the oxidized and reduced forms of DsbDγ reflect its structural stiffness. Another interesting feature of DsbDγ is the unusually high pK_a value of 9.3 for the exposed active site Cys461: this value, determined by two different methods, surprisingly turned out to be some 2.5 units higher than expected on the basis of the redox potential of this protein (-235mV) [180, 181]. Therefore, DsbDγ appears to be the first thioredoxin-like domain where the exposed active site cysteine residue exhibits a pK_a that is not lower than, but rather on a par with that of a normal cysteine side-chain. Overall, the above results highlighted two important things about the γ domain of DsbD: its unusual redox properties and its extreme rigidity. These features are likely to play a role in the functioning of DsbDγ as a covalently linked electron shuttle between the membrane domain and the N-terminal periplasmic domain of DsbD [180].

2.3.6. How Does DsbD Work?

An elegant approach to tracing the pathway of electrons through the DsbD molecule was developed by Katzen and Beckwith [178]. To test the possibility that disulfide bond oxidation and reduction constitutes the mechanism of electron transfer in DsbD, they cleaved DsbD into its three structural domains, each containing two essential cysteines. When coexpressed, these truncated proteins restored DsbD function, thus ensuring electron transfer from the cytoplasm to the periplasm. DsbD was split at two different positions: (1) between the N-terminal periplasmic domain and the hydrophobic core and (2) between the hydrophobic core and the thioredoxin fold. In this way, constructs were generated encoding five different polypeptides, designated α, β, γ, αβ, and βγ. Plasmids expressing DsbD derivatives were introduced into a *dsbD* null mutant, and their ability to reduce DsbC was assayed. Nearly all DsbC from the strain expressing wild type DsbD maintained its active site cysteine in the reduced state. However, when any of the three DsbD structural domains was removed from the system, DsbC would accumulate in the oxidized state. These results indicated that all three domains of DsbD are required, but do not need to be part of a single polypeptide chain, for the transfer of reducing potential from the cytoplasm to DsbC [178].

Electrons Are Sequentially Transferred from One Domain to Another

The simplest way to establish the sequential order in which different domains participate in the reaction is to disrupt the electron pathway by taking one domain away from the system, and then to assess the redox status of the remaining components. Those domains that come into action at a later step than the removed component should be unable to receive reducing potential, and thus would remain oxidized. In contrast, those domains that accomplish their function at a step prior to that of the removed component would accumulate in the reduced form or, alternatively, their redox state would remain unchanged. Under the experimental conditions employed in the above study, reduced and oxidized forms of α, β, and γ were clearly distinguishable [178].

According to data obtained in that series of experiments, in the process of electron transfer from cytoplasmic thioredoxin through DsbD, γ acts before α but later than β. When α is removed from the three-component system, the redox state of the remaining peptides remains unchanged. This finding makes it reasonable to assume that α is the last destination of the electrons in DsbD before they are transferred to periplasmic substrates. In the absence of β the other components become completely oxidized, implying that β is involved in the earliest step of transferring electrons from thioredoxin. In sum, these results strongly suggest an electron pathway proceeding as follows:

$$\text{thioredoxin} \to \beta \to \gamma \to \alpha \to \text{DsbC};$$

they are also consistent with the notion that electron flow within DsbD occurs via a succession of disulfide exchange reactions [178].

The above model, proposed by Katzen and Beckwith for the mechanism of DsbD action, was tested by Bardwell and his collaborators, who performed a reconstitution of DsbD activity by biochemical methods [182]. Different polypeptides corresponding to the α, β, γ, and βγ domains were purified to homogeneity. Using these domains, the authors could reconstitute DsbD activity, and, for the first time, reproduce *in vitro* the electron transfer pathway from

NADPH and thioredoxin to DsbC and DsbG. Their data clearly showed that electrons flow from NADPH to thioredoxin, then successively to the β, γ, and α domains and to DsbC or DsbG, as anticipated by the model [178]. To find out whether the electron flow is thermodynamically driven, the redox potentials of the γ and α domains were determined and found to be -241mV and -229mV, respectively [182]. Given that the redox potentials of thioredoxin and of DsbC had been measured earlier respectively at -270mV and -159mV, [183], the redox potential appeared to increase as electrons flowed down the isomerization pathway from NADPH to DsbC, as shown below, and the direction of electron flow within DsbD was thus thermodynamically driven.

	NADPH →	Thioredoxin →	γ →	α →	DsbC
Redox Potential	-315mV	-270mV	-241mV	-229mV	-159mV

In Figure 12, successive steps in the pathway of electron flow from reduced thioredoxin to DsbC are shown schematically to illustrate the participation of different pairs of cysteine residues in each step.

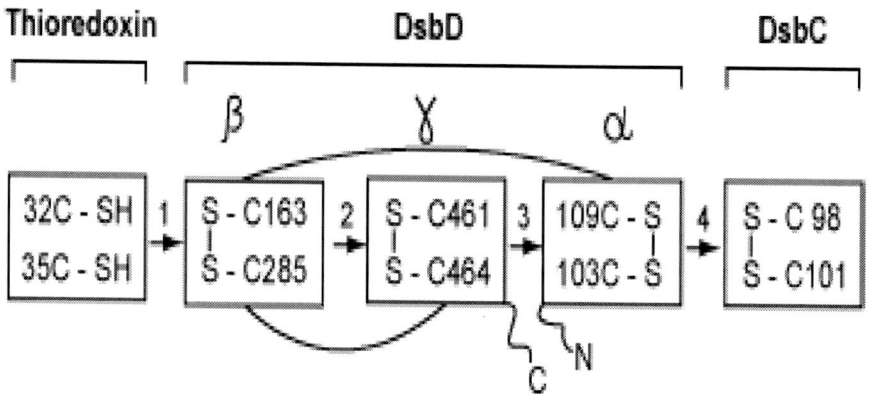

Figure 12. Schematic representation of successive steps on the pathway of DsbD-catalyzed electron flow from reduced thioredoxin to DsbC. The numbers of the cysteine residues involved in thiol-disulfide exchange reactions are indicated. The arrows mark steps 1-4. Cys32 of reduced thioredoxin interacts with Cys163 of the DsbD β-domain (step1). This presumably resolves the disulfide between Cys163 and Cys285 of DsbD. In step 2, Cys163 and Cys285 presumably donate electrons to the dithiols at C461 and C464 of the DsbD γ domain. In step 3, the reduced dithiol of C461 attacks the disulfide at C103 and Cys109 of the DsbD α domain, which transfer electrons to the oxidized dithiol motif of DsbC (step 4). Adapted from Goulding et al. [171].

It appears that electrons are delivered specifically to the active site disulfide bonds of oxidized DsbC and DsbG, avoiding the many other disulfide bonds in the periplasm. How is this specificity achieved? To get an insight into the structural basis of specific interactions between electron donors and acceptors in each successive step of the electron transfer, one would need to examine in each case the structure of a trapped complex of two partners forming a mixed disulfide. In the subsequent paragraphs, we shall consider recent data which contribute to a better understanding of these effects.

Step 1. Electron Transfer from Reduced Thioredoxin to the β Domain of DsbD

The specific mechanism whereby DsbDβ mediates electron transfer from the cytoplasm across the membrane was examined in a study by Krupp et al. [184]. Reduced thioredoxin (thioredoxin A, TrxA) is a small cytoplasmic protein (~ 10 kDa) that contains a single redox active site Cys32-Gly-Pro-Cys35 [185]. It was shown that *E. coli* lacking thioredoxin accumulates oxidized DsbD, and that both cysteines of the dithiol motif of TrxA are required for electron transfer. A substitution of TrxA-C35S but not TrxA-C32S produced a whole spectrum of disulfide-linked intermolecular intermediates. Hence, a mutant thioredoxin with a substituted second cysteine residue was capable of attacking numerous cytoplasmic disulfides but unable to resolve mixed disulfide intermediates. *E coli* cells expressing TrxA-C35S and DsbD variants with single cysteine to alanine substitutions were studied by immunoblotting under the conditions when all cellular disulfide exchange was quenched. Analysis of mixed disulfides detected in these experiments provided a strong evidence that electron transfer between thioredoxin and DsbD involves the formation of a mixed disulfide between Cys32 of thioredoxin and Cys163 of DsbDβ [184].

Step 2. Electron Transfer from Cys 285 of the β Domain to the γ Domain of DsbD

Assuming that electron transfer from thioredoxin to the periplasmic domain of DsbD involves a thiol-disulfide exchange between the Cys163-Cys285 pair of DsbDβ and the Cys461- Cys464 pair of DsbDγ, one has to explain how a mixed disulfide can form between these two cysteine pairs. Given that both Cys163 and Cys285 face the cytoplasmic side of the membrane [177], (see above), it remains to be determined in what way DsbDγ, located in the periplasm, is able access them from the other side of the membrane, across a distance of about 60Å. Katzen and Beckwith have proposed several ways in which it can be accomplished. In one scenario, the entire γ domain or just its helix II and the loop containing the two cysteines [179], is imported across or inserted into the membrane, thereby allowing thiol-disulfide bond exchange between the two cysteine pairs.

Another scenario involves a major rearrangement of the hydrophobic core which allows thiol-disulfide bond exchange to proceed in the periplasmic space. Under this scenario, the transmembrane domain of DsbD might adopt a funnel-like shape, with the "funnel" opening toward the cytoplasm when the Cys163-Cys285 disulfide is formed. Direct oxidation of the Cys163-Cys285 pair by DsbDγ is thus only conceivable when a structural rearrangement occurs in the transmembrane domain such that the "funnel" opens toward the periplasm when the Cys163-Cys285 disulfide is reduced by cytoplasmic thioredoxin. An third, a protein-bound cofactor may assist electron transfer from the Cys163-Cys285 pair to the Cys461-Cys464 disulfide bond of DsbDγ [177].

Step 3. Electron Transfer from DsbDγ to DsbDα

As described above, the recently determined high-resolution crystal structure of DsbDγ revealed that this domain is very rigid and undergoes hardly any conformational change upon disulfide reduction or interaction with DsbDα. This structural rigidity is consistent with a model in which a rigid DsbDγ molecule acts as a "stiff electron shuttle": it is reduced by docking to a site in the DsbDβ transmembrane domain and then, without relevant conformational changes, transfers electrons to the structurally adaptable N-terminal domain, DsbDα [180]. Rozhkova et al. have demonstrated that Cys109 of DsbDα and Cys461 of DsbDγ form a mixed disulfide bond, and determined the X-ray structure of the resulting complex [181].

The structure of this mixed disulfide termed DsbDα-SS-DsbDγ reveals in detail how the N-terminal immunoglobulin-like domain of DsbD interacts with the C-terminal thioredoxin-like domain. As described above (Figure 11), Cys109 in isolated oxidized DsbDα is shielded by the so-called cap-loop region (segment Asp68-Gly72) [171]. In a DsbDα-SS-DsbDγ complex, Cys109 of DsbDα becomes accessible due to a conformational switch of that region, which results in the opening of the cap. A surface representation of DsbDα-SS-DsbDγ shows a relatively planar region between the C-terminus of DsbDα and the N-terminus of DsbDγ, which is supposed to be oriented towards the surface of the membrane. Besides the interdomain disulfide bond, the DsbDα -SS-DsbDγ domain interface is characterized by only a limited number of specific interactions, namely several hydrogen bonds. Overall, the number of specific contacts is rather limited relative to the size of the interface area (1301 $Å^2$). In a model of DsbDα-SS-DsbDγ complex proposed by Kim et al., DsbDγ forms a complementary interface with the DsbDα molecule whereby no noticeable clashes between residues of DsbDγ and DsbDα exist, indicating that DsbDγ can bind DsbDα without major conformational rearrangement [179].

Step 4. Electron Transfer from DsbDα to DsbC

Experiments involving the expression of DsbD mutants with systematically mutated cysteines revealed the formation of a mixed disulfide between Cys98 of DsbC and Cys109 of DsbDα [184]. The latter complex could be observed in cells expressing the mutants DsbC C101A and DsbD C103A. These mutants lack local free cysteines able to resolve the intermolecular disulfide bond that forms between DsbDα Cys109 and DsbC Cys98. To investigate the interactions between DsbC and DsbD, Goldstone et al. [186] expressed and purified DsbDα and DsbC. The results of experiments involving sulfhydryl labeling of mixtures of these two proteins in defined redox states, combined with protein disulfide isomerase assays, showed that DsbDα can reduce and fully activate oxidized DsbC. Further analysis indicated that a single DsbDα molecule would bind a DsbC dimer. No complex formation was observed when the thioredoxin domain of DsbC alone was substituted for full-length DsbC.

This finding has aroused considerable interest since it implies that regions of DsbC distant from the active site are involved in the formation of the complex, and suggests that DsbDα – DsbC interaction may be specific in nature. To test the validity of such a suggestion, the crystal structure of a trapped DsbC- DsbDα complex was solved, and interactions between the two proteins were examined in detail [172]. As had been expected, DsbDα binds into the central cleft of the V-shaped DsbC homodimer, which assumes a closed state upon formation of the complex. The most interesting fact is that DsbDα engages both active sites of DsbC, forming two asymmetric binding sites.

The primary binding site involves interactions between residues surrounding the active sites of both DsbDα and DsbC, and contains the two catalytic cysteine residues, DsbDα's Cys109 and DsbC's Cys98, which form a disulfide bond between the two molecules (Figure 13). In the case of DsbDα, the primary binding region lies entirely in the catalytic subdomain, and no primary binding contacts are made with residues in the core of the DsbDα Ig fold. The secondary binding site includes contacts between the active site region of the second DsbC monomer and residues from DsbDα's Ig fold. The recognition of both DsbC active sites explains why DsbDα selectively forms a complex with full-length dimeric DsbC, but not with the truncated catalytic domain, which is monomeric in solution.

The following model is being suggested for selective activation of oxidized dimeric DsbC [172]. The electron transport reaction is initiated by the binding of reduced DsbDα into the substrate binding cleft of oxidized DsbC. The binding induces major conformational changes, and DsbC assumes a closed conformation allowing for interactions between the two DsbC catalytic domains and DsbDα. The interaction with both catalytic domains explains the preferential binding of DsbDα to the dimeric form of DsbC. A comparison of the oxidized DsbDα structure (Figure 11) with the structure of the DsbDα-DsbC complex reveals significant conformational changes in the cap regulating the accessibility of the DsbDα active site. The open conformation of the DsbDα active site cap, existing in the reduced DsbDα structure, exposes the DsbC binding pocket and the catalytic residue Cys109 of DsbDα.

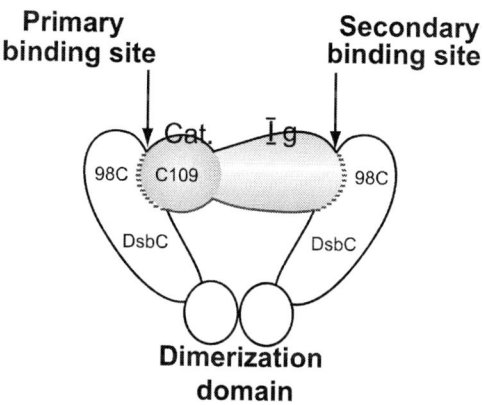

Figure 13. Asymmetric structure of the (DsbC) · (α domain of DsbD) complex. Cat. and Ig, catalytic and Ig subdomain of the DsbD α domain, respectively. The α domain of DsbD binds into the central cleft of DsbC, interacting with both active sites of DsbC. However, these two binding sites are non-identical. The primary binding site involves interactions between residues surrounding the active sites of both DsbD α domain and DsbC. It contains two catalytic cysteine residues, the DsbC Cys98 and the α-domain Cys109, which can form a disulfide bond between the two molecules. In the case of DsbD α-domain, the primary binding site region lies entirely within the catalytic subdomain. The secondary binding site includes contacts between the active site region of the second DsbC monomer (namely, Cys98), and residues from the Ig fold.

Available information suggests that the four sulfur atoms of the catalytic residues involved in the thiol-disulfide exchange reaction are aligned appropriately for the subsequent nucleophilic attack of DsbDα Cys109 on the active site disulfide of oxidized DsbC. In the course of the reaction, Cys109 forms an intermolecular disulfide bond with Cys98 of DsbC. The intermediate disulfide bond is resolved by the nucleophilic attack of the free Cys103 from DsbDα, leaving DsbC reduced and DsbDα oxidized. After DsbC activation and DsbDα oxidation, the cap moves into a "closed" position, shielding the DsbDα active site.

Finally, the DsbC- DsbDα complex dissociates due to charge repulsion between the DsbDα residue Asp68 (see Figure 11) and the Cys98 thiolate of reduced DsbC [172].

Taken together, the structural properties of DsbDα provide a clue to understanding how this thiol-disulfide oxidoreductase is able to select its substrates. Namely, it transfers electrons to DsbC but not to DsbA, even though DsbC and DsbA have similar structural features. Special about DsbDα is the fact that electron transfer is mediated by an immunoglobulin-like fold. The cysteines in this fold are less promiscuous than those in the thioredoxin fold, as no mixed disulfides have been observed between DsbDα and random proteins containing dithiol-disulfide groups [184]. It appears that the DsbDα's immunoglobulin-like fold is a variation of a common fold that has developed a relatively new electron-transporting function, perhaps uniquely tailored to provide reducing equivalents for thiol-disulfide exchangers coupled to protein folding in bacterial cells. DsbDα's immunoglobulin-like structure is distinguished by the presence of a cap region, whose suggested purpose is to protect the active site from illegitimate redox reactions [171]. And finally, the unique way of binding DsbDα within the cleft separating two DsbC monomers, whereby DsbDα not only does not interact with residues around the active site disulfide of a DsbC subunit, but forms specific, additional contacts with the second subunit in the DsbC dimer, ensures highly specific recognition of its target protein.

Conclusion

Thiol/disulfide oxidoreductases catalyze the formation and isomerization of disulfide bonds in proteins, i.e. those reactions that serve as rate-limiting steps in the folding of certain (preliminary secreted) proteins. In eukaryotic cells, these reactions occur in the endoplasmic reticulum and are catalyzed by a single enzyme (PDI), while in the cells of Gram-negative bacteria formation and isomerization of disulfide bonds take place in the periplasmic space and are catalyzed by two separate enzymes. Both oxidation and isomerization are necessary for allowing the full complement of native disulfide bond formation. All thiol/disulfide oxidoreductases utilize a common catalytic mechanism based on thiol-disulfide exchange reactions between an active site catalytic disulfide and cysteine residues of a substrate protein. Oxidation involves the transfer of an active site disulfide to substrate proteins, while isomerization requires the active site cysteines to be in a reduced form so that they can attack non-native disulfides in substrate proteins, thereby catalyzing their rearrangement.

The best understood system of disulfide bond formation in folding proteins occurs in the periplasm of Gram-negative bacteria. In *E. coli,* the formation of protein disulfide bonds depends on the products of *dsb* genes which contain active site disulfides and are members of the thioredoxin protein superfamily. Five Dsb proteins: DsbA, DsbB, DsbC, DsbD, and DsbG, are involved in disulfide bond formation and rearrangement. The addition of disulfide bonds within the periplasm is catalyzed by the soluble periplasmic protein DsbA. The active site of DsbA is regenerated by the transfer of electrons to the integral membrane protein DsbB which, in turn, passes electrons to quinone and the electron transport chain.

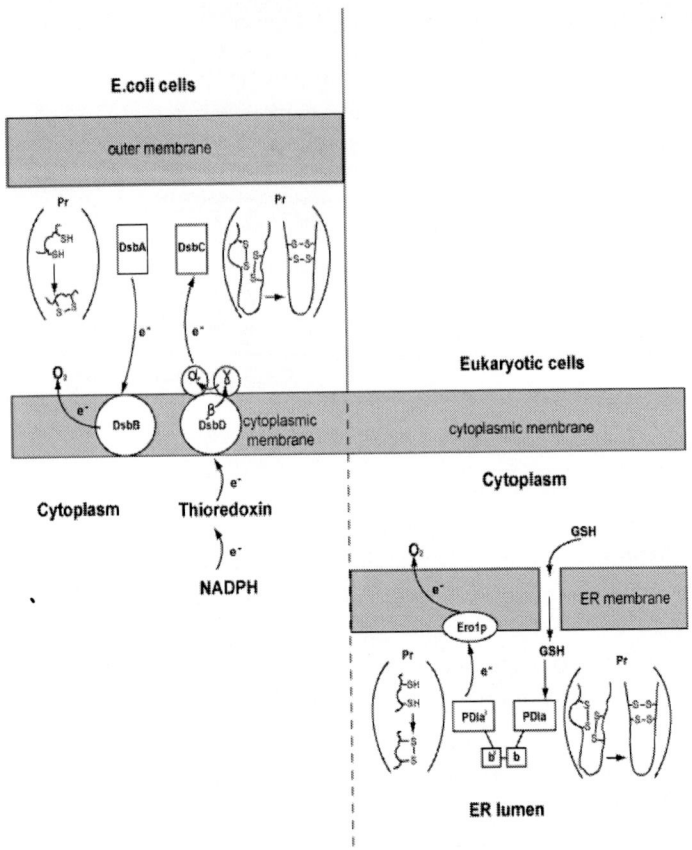

Figure 14. Left: schematic representation of the oxidation and isomerization pathways in *E. coli* periplasm. In the oxidation pathway, reduced proteins are oxidized by the disulfide bond at the active site of DsbA, a rapid and unidirectional process which helps catalyze their folding. DsbA is reoxidized by the inner membrane protein DsbB, which passes electrons to oxygen. In the isomerization pathway, DsbC is kept in the reduced form by the membrane protein DsbD. The proposed direction of electron flow within the DsbD domains α, β and γ is shown by curved arrows. DsbD is kept reduced by the reducing power of thioredoxin, which is reduced by thioredoxin reductase and then by NADPH. Right: schematic representation of the oxidation and isomerization pathways in the endoplasmic reticulum of eukaryotic cells. In the oxidation pathway, the reduced PDI active site cysteine pair passes electrons to the membrane-bound protein Ero1p, which transfers electrons to oxygen. In the isomerization pathway, reduced glutathione (GSH), coming from the cytoplasm, keeps a proportion of PDI active sites in the reduced state and competent for disulfide bond rearrangement. a, b, b' and a' designate PDI domains, only the a and a' domains being catalytically active. Pr, substrate protein; ER, endoplasmic reticulum.

Incorrect disulfide bonds formed by DsbA may trap proteins in non-native conformations. Disulfide bond isomerases DsbC and DsbG facilitate the folding of proteins with multiple disulfide bonds. Because of the oxidative environment of the periplasm, the isomerization pathway of *E. coli* requires additional proteins to maintain DsbC and DsbG in the reduced state. In a striking parallel to the DsbB-DsbA system, DsbC is dependent on the cytoplasmic membrane protein DsbD for the regeneration of its reduced active site. However, in contrast to the DsbB-DsbA pathway, which uses intramembranous electron transfer components, DsbD is involved in electron transfer with cytoplasmic proteins. The cytoplasmic thioredoxin pathway passes electrons to DsbD, which keeps DsbC in the reduced state and competent for disulfide bond rearrangement.

Figure 14, left illustrates the situation in the bacterial periplasm, where DsbA is kept oxidized in order to function as a disulfide donor, while DsbC is kept reduced so it can work as a disulfide isomerase.

The coexistence of these two paths within the same periplasmic compartment is only possible if specific mechanisms prevent futile cycles of DsbB oxidizing DsbC while DsbD reduces DsbC, or DsbD reducing DsbA while DsbB oxidizes DsbA. Such mechanisms do exist; they are based on the ability of DsbB and DsbD to discriminate between DsbA and DsbC. Thus, DsbB cannot oxidize DsbC because DsbC active site cysteines are protected by dimerization, whereas the active site cysteines of monomeric DsbA are not. By contrast, DsbD (i.e. its periplasmic α-domain) can only interact with the dimeric form of DsbC, forming a set of specific contacts that cannot be formed with monomeric DsbA.

The pathway of protein disulfide bond formation in the bacterial periplasm provides a useful analogy to the protein oxidation system in eukaryotic endoplasmic reticulum (Figure 14, right). In both cases, disulfide bonds are introduced into folding proteins by transfer from a disulfide bond carrier that is a member of the thioredoxin family (DsbA in prokaryotes and PDI in eukaryotes). Those disulfide bond carriers, in turn, receive disulfide bonds from a different class of thiol-oxidoreductases, proteins that couple intracellular biochemical reactions producing oxidizing potential with the formation of protein disulfide bonds. In both prokaryotic and eukaryotic systems, this function is fulfilled by membrane-associated oxidoreductases (DsbB and flavoenzyme Ero1p, respectively). The mechanisms they use are obviously similar: each of them has two pair of essential cysteine residues, one located at the active center and the other placed in a flexible loop which protrudes into the periplasm (or into the lumen of the endoplasmic reticulum) and engages directly in thiol-disulfide exchange with DsbA (or PDI). Further, electrons are transferred to the second cysteine pair and finally, via pathways which are different in prokaryotes and eukaryotes, to oxygen (Figure 14).

Eukaryotic PDI is both structurally and functionally more complex than DsbA or DsbC. In contrast to these enzymes, which contain just one active site per monomer, the PDI monomer has a multidomain structure consisting of 4 thioredoxin-like domains (Figure 14, right). Both the a and a' domains contain a redox-active CXXC motif, whereas the b and b' domains do not. Isomerization of complex substrates, particularly ones in which substantial conformational rearrangements must occur, requires all four of the thioredoxin-like domains. A fraction of PDI has to be kept active as an isomerase; this is achieved by means of reduced glutathione, which is imported into the endoplasmic reticulum from the cytosol. Figure 14, right represents a hypothetic intercompartmental pathway which would allow cytosolic GSH access to the endoplasmic reticulum. Fig 14, left shows that bacterial cells possess another isomerization pathway, the cytoplasmic thioredoxin pathway, which passes electrons to DsbD and then to DsbC.

The ability of PDI to function both as a disulfide isomerase and a dithiol oxidase raises the question of how it is capable of balancing the opposing redox processes of disulfide bond oxidation and isomerization that are required to achieve optimal protein folding. The next major breakthrough in studying the mechanisms of PDI action will come with the resolution of this crucial issue. Available data indicate that the active sites of PDI are asymmetric. Despite the high overall structural similarity, a few significant differences exist between the a and a' active sites, suggesting that the a' domain is a better oxidant that the a domain. What is the role of functional non-equivalence of the PDI's active centers? This question attracts considerable attention and remains to be answered.

Another interesting topic for future research will be to determine the mechanisms whereby the electrons required for oxidative and reductive processes are passed into or through membranes. For instance, in the case of the isomerization pathway in bacteria, electrons flowing from the cytoplasm to the cell envelope may be using several transfer steps between proteins, which remain unexplored. How are those electrons transferred from thioredoxin on the cytoplasmic side of the membrane to DsbD, and thence to the periplasmic side of the same membrane protein? These are the kind of questions that will pose a challenge to future researchers seeking to understand the intricate mechanisms elaborated by the cell for the oxidation of cysteine residues and the isomerization of disulfide bonds.

References

[1] Anfinsen, C.B., Haber, E., Sela, M., and White, F.H. 1961, *Proc. Natl. Acad. Sci. USA*, 47, 1309.
[2] Goldberger, R.F., Epstein, C.J., and Anfinsen, C.B. 1963, *J. Biol. Chem.*, 238, 628.
[3] Givol, D., DeLorenzo, F., Goldberger, R.F., and Anfinsen, C.B. 1965, *Proc. Natl. Acad. Sci. USA,* 53, 676.
[4] Creighton, T.E., Hillson, D.A., and Freedman, R.B. 1980, *J. Mol. Biol.*, 142, 43.
[5] Koivu, J., and Myllyla, R. 1987, *J. Biol. Chem.*, 262, 6159.
[6] Bulleid, N.J., and Freedman, R.B., 1988, *Nature*, 335, 649.
[7] Freedman, R.B. 1984, *Trends Biochem. Sci.*, 9, 438.
[8] Freedman, R.B., Hirst, T.R., and Tuite, M.F., 1994, *Trends Biochem. Sci.*, 1994, 19, 331.
[9] Yao, Y., Zhou, Y.-c., and Wang, C.-c. 1997, *EMBO J.*, 16, 651.
[10] Vinci, F., Catharino, S., Frey, S., Buchner, J., Marino, G., Pucci, P., and Ruppolo, M. 2004, *J. Biol. Chem.* 279, 15, 15059.
[11] LaMantia, M.L., Miura, T., Tachikawa, H., Kaplan, H.A., Lennharz, W.J., and Mizunaga, T. 1991, *Proc. Natl. Acad. Sci. USA*, 88, 4453.
[12] LaMantia, M.L., and Lennharz, W.J. 1993, *Cell*, 74, 899.
[13] Tian, G., Xiang, S., Noiva, R., Lennarz, W., and Schindelin, H. 2006, *Cell*, 124, 61.
[14] Bardwell, J.C., McGovern, K., and Beckwith, J. 1991, *Cell*, 67, 581.
[15] Ohba, H., Harano, T., and Omura, T. 1981, *J. Biochem.*(Tokyo), 89, 889.

[16] Tojo, H., Asano, T., Kato, K., Udaka, S., Horiuchi, R., and Kahinuma, A. 1994, *J. Biotechnol.*, 33, 55.
[17] Solovyov, A., and Gilbert, H.F. 2004, *Protein Sci.*, 13, 1902.
[18] Locker, J.K., and Griffiths, G. 1999, *J. Cell. Biol.*, 144, 267.
[19] Hwang, C., Sinskey, A.J., and Lodish, H.F. 1992, *Science*, 257, 1496.
[20] Frand, A., Cuozzo, J.W., and Kaiser, C.A. 2000, *Trends. Cell. Biol.*, 10, 203.
[21] Noiva, R. 1994, *Prot. Expr. Purif.*, 5, 1.
[22] Gilbert, H.F. 1998, *Methods Enzymol.*, 290, 26.
[23] Helenius, A. 2001, *Phil. Trans.R. Soc.Lond.B Biol. Sci.*, 356, 147.
[24] Huth, J.R., Perini, F., Lockridge, O., Bedows, E., and Ruddon,R.W. 1993, *J. Biol. Chem.*, 268, 16472.
[25] Weissman, J.S., and Kim, P.S. 1993, *Nature*, 365, 185.
[26] Edman, J.C., Ellis, L., Blacher, R.W., Roth, R.A., and Rutter, W.J. 1985, *Nature* (London), 317, 267.
[27] Ramakrishna Kurup, C.K., Raman, T.S., and Ramasarma, T. 1966, *Biochim. Biophys. Acta*, 113, 255.
[28] Freedman, R.B. 1980, *Biochem. J.,* 191, 373.
[29] Holmgren, A. 1985, *Annu. Rev. Biochem.*, 54, 237.
[30] •Darby, N.J., Kemmink, J., and Creighton, T.E. 1996, *Biochemistry*, 35, 10517.
[31] Darby, N.J., van Straaten, M., Penka, E., Vincentelli, R., and Kemmink, J. 1999, *FEBS Lett.*, 448, 167.
[32] Kemmink, J., Darby, N.J., Dijkstra, K., Nilges, M., and Creighton, T.E. 1997, *Curr. Biol.*, 7, 241.
[33] Koivunen, P., Pirneskovski, A., Karvonen,P., Ljung, J., Helaakoski, T., Notbohm, H., and Kivirikko, K.I. 1999, *EMBO J.*, 18, 65.
[34] Kemmink, J., Darby, N., Dijkstra, K., Nilges, M., and Greighton, T.E. 1996, *Biochemistry*, 35, 7684.
[35] Martin, J.L. 1995, Structure, 3, 245.
[36] Vuori, K., Myllyla, R., Pihlajaniemi, T., and Kivirikko, K.I. 1992, *J. Biol. Chem.*, 267, 7211.
[37] Darby, N.J., and Greighton, T.E. 1995, *Biochemistry*, 34, 11725.
[38] Doolittle, F.F. 1995, *Annu. Rev. Biochem.*, 64, 287.
[39] Klappa, P., Ruddock, L.W., Darby, N.J., and Freedman, R.B. 1998, *EMBO J.*, 17, 927.
[40] Pirneskovski, A., Klappa, P., Lobell, M., Williamson, R.A., Byrne, L., Alanen, H.I., Salo, K.E.H., Kivirikko, K.I., Freedman, R.B., and Ruddock, L.W.2004, *J. Biol. Chem.*, 279, 10374.

[41] Darby, N.J., Penka, E., and Vincentelli, R. 1998, *J. Mol. Biol.*, 276, 239.
[42] Hawkins, H.C., and Freedman, R.B., 1991, *Biochem. J.*, 275, 335.
[43] Kallis, G.-B., and Holmgren, A. 1980, *J. Biol. Chem.*, 255, 10261.
[44] Nelson, J.W., and Greighton, T.E. 1994, *Biochemistry*, 33, 5974.
[45] Kortemme, T., and Creghton, T.E. 1995, *J. Mol. Biol.*, 253, 799.
[46] Lappi, A.K.,, Lensink, M.F., Alanen,H.I., Salo, K.E.H., Lobell, M., Juffer, A.H., and Ruddock, L.W. 2004, *J. Mol. Biol.*, 335, 283.
[47] Xiao, R., Wilkinson, B., Solovyov, A., Winther, J.R., Holmgren, A., Lundstrom-Ljung, J., and Gilbert, H.F. 2004, *J. Biol. Chem.*, 279, 49780.
[48] Wilkinson, B., Xiao, R., and Gilbert, H.F. 2005, *J. Biol. Chem.*, 280, 11483.
[49] Rothwarf, D.M., Li, Y.J., and Scheraga, H.A. 1998, *Biochemistry*, 37, 3760.
[50] Creighton, T.E., and Goldenberg, D.P. 1984, *J. Mol. Biol.*,179, 497.
[51] Creighton, T.E. 1988, *BioRssays*, 8, 57.
[52] Kim, P.S., and Baldwin, R.L., 1990, *Annu. Rev. Biochem.*, 59, 631.
[53] Schwaller, M., Wilkinsoin, B., and Gilbert, H.F. 2003, *J. Biol. Chem.*, 278, 7154.
[54] Farquhar, R., Honey, N., Murant, S.J., Boissier, P., Schultz, L., Montgomery, D., Ellis, R.W., Freedman, R.B., and Tuite, M.F. 1991, *Gene*, 108, 81.
[55] Woycechowsky, K.J., and Raines, R.T. 2000, *Curr. Opin. Chem. Biol.*, 4, 533.
[56] Laboissiere, M.C.A., Sturley, S.L.and Raines, R.T. 1995, *J. Biol. Chem.*, 270, 28006.
[57] Kersteen, E.A., and Raines, R.T. 2003, *Antiox. and Redox Sign.* 5, 413.
[58] Scherens, B., Dubois, E., and Messenguy, F. 1991, *Yeast*, 7, 185.
[59] Walker, K.W., Lyles, M.M., and Gilbert, H.F. 1996, *Biochemistry*, 35, 1972.
[60] Walker, K.W., and Gilbert, H.F. 1997, *J. Biol. Chem.*, 272, 8845.
[61] Kerṣteen, E.A., Barrows, S.R., and Raines, R.T. 2005, *Biochemistry*, 44, 12168.
[62] McCarthy, A.A., Haebel, P.W., Toronen, A., Rybin, V., Baker, E.N., and Metcalf, P. 2000, *Nat. Struct. Biol.*, 7, 196.
[63] Sun, X.X., and Wang, C.C. 2000, *J. Biol. Chem.*, 275, 22743.
[64] Zheng, J., and Gilbert, H.F. 2001, *J. Biol. Chem.*, 276, 15747.
[65] Lyles, M.M., and Gilbert, H.F. 1991, *Biochemistry*, 30, 613.
[66] Frand, A.R., and Kaiser, C.A. 1998, *Mol. Cell*, 1, 161.
[67] Pollard, G., Travers, K.J., and Weissman, J.S. 1998, *Mol. Cell*, 1, 171.

[68] Cabibbo, A., Pagani, M., Fabbri, M., Rocchi, M., Farmery, M.R., Bulleid, N.J., and Sitia, R. 2000, *J. Biol. Chem.*, 275, 4827.
[69] Pagani, M., Fabbri, M., Benedetti, C., Fassio, A., Pilati, S., Bulleid, N.J., Cabibbo, A., and Sitia, R. 2000, *J. Biol. Chem.*, 275, 23685.
[70] Bentham, A.M., Cabibbo, A., Fassio, A., Bulleid, N.J., Sitia, R., and Braakman, I. 2000, *EMBO J.*, 19, 4493.
[71] Mezghrani, A., Fassio, A., Bentham, A.M., Simmen, T., Braakman, I., and Sitia, R. 2001, *EMBO J.*, 209, 6288.
[72] Pagani, M., Pilati, S., Bertoli, G., Valsasina, B., and Sitia, R. 2001, *FEBS Lett.*, 508, 117-120.
[73] Frand, A.R., and Kaiser, C.A. 1999, *Mol. Cell*, 4, 469.
[74] Frand, A.R., and Kaiser, C.A. 2000, *Mol. Biol. Cell*, 11, 2833.
[75] Frand, A.R., Cuozzo, J., and Kaiser, C.A. 2000, *Trends Cell Biol.*, 5, 203.
[76] Tu, B.P., Ho-Schleyer, S.C., Travers, K.J., and Weissman, J.S. 2000, *Science*, 290, 1571.
[77] Tu, B.P., and Weissman, J.S. 2002, *Mol. Cell*, 10, 983.
[78] Tu, B.P., and Weissman, J.S. 2004, *J. Cell. Biol.*, 164, 341.
[79] Gilbert, H.F. 1990, *Adv. Enzymol. Relat. Areas Mol. Biol.*, 63, 69.
[80] Gliszczynska, A., and Koziolowa, A. 1998, *J. Chromatogr.* A. 822, 59.
[81] Gross, E., Sevier, C.S., Heldman, N., Vitu, E., Bentzur, M., Kaiser, C.A., Thorpe, C., and Fass, D. 2006, *Proc. Natl. Acad. Sci. USA*, 103, 299.
[82] Gross, E., Kastner, D.B., Kaiser, C.A., and Fass, D. 2004, *Cell*, 117, 601.
[83] Bardwell, J.C.A. 2004, *Nat. Struct. Mol. Biol.*, 11, 582.
[84] Hiniker, A., and Bardwell, J.C.A. 2004, *Trends Biochem. Sci.*, 29, 516.
[85] Lyles, M.M., and Gilbert, H.F. 1994, *J. Biol. Chem.*, 269, 30946.
[86] Westphal, V., Darby, N.J. and Winther, J.R. 1999, *J. Mol. Biol.*, 286, 1229.
[87] Norgaard, P., and Winther, J.R. 2001, *Biochem. J.*, 356, 269.
[88] Holst, B., Tachibana, C., and Winther, J.R. 1997, *J. Cell Biol.*, 138, 1229.
[89] Lu, X., Gilbert, H.F., and Harper, W.J. 1992, *Biochemistry*, 31, 4205.
[90] Tsai, B., and Rapoport, T.A. 2002, *J. Cell Biol.*, 159, 207.
[91] Kulp, M.S., Frickel, E.-M., Ellgaard, L., and Weissman, J.S. 2006, *J. Biol.Chem.*, 281, 876.
[92] Bader, M., Whinter, J.R., and Bardwell, J.C.A. 1999, *Nat.Cell. Biol.*, 1, E56.
[93] Cuozzo, J.W., and Kaiser, C.A., 1999, *Nature Cell Biol.* 1, 130.
[94] Ohtake, Y., and Yabuuchi, S. 1991, *Yeast*, 7, 953.
[95] Chakravarthi, S., and Bulleid, M.N.J. 2004, *J. Biol. Chem.*, 279, 39872.
[96] Molteni, S.N., Fassio, A., Ciriolo, M.R., Filomeni, G., Pasqualetto, E., Fagioli, C., and Sitia, R. 2004, *J. Biol. Chem.*, 279, 32667.

[97] Jessop, C.E., and Bulleid, N.J. 2004, *J. Biol. Chem.*, 279, 55341.
[98] Bass, R., Ruddock, L.W., Klappa, P., and Freedman, R.B. 2004, *J. Biol. Chem.*, 279, 5257.
[99] Frickel, E.-M., Frei, P., Bouvier, M., Stafford, W.F., Helenius, A., Glockshuber, R., and Ellgaard, L. 2004, *J. Biol. Chem.*, 279, 18277.
[100] Fratelli, M., Demol, H., Puype, M., Casagrande, S., Eberini, I., Salmona, M., Bonetto, V., Mengozzi, M., Duffieux, F., Miclet, E., Bachi, A., Vandekerck-hove, J., Gianazza, E., and Ghezzi, P. 2002, *Proc. Natl. Acad. Sci. USA*, 99, 3505.
[101] Banhegy, G., Lusini, L., Puskas, F., Rossi, B., Fulceri, R., Braun, L., Mile, V., di Simplicio, P., Mandl, J., and Benedetti, A. 1999, *J. Biol. Chem.*, 274, 12213.
[102] Le Gall, S., Neuhof, A., and Rapoport, T. 2004, *Mol. Biol. Cell*, 15, 447.
[103] Tu, B.P., Ho-Schleyer, S.C., Travers, K.J., and Weissman. J.S. 2002, *Arch. Biochem. Biophys.*, 405, 1.
[104] Wolin, S.L. 1994, Cell, 77, 787.
[105] Tan, J.T., and Bardwell, J.C.A. 2004, *ChemBiochem.*, 5, 1479.
[106] Collet, J.-F., and Bardwell, J.C.A. 2002, *Mol. Microbiol.*, 44, 1.
[107] Kamitani, S., Akiyama, Y., and Ito, K. 1992, *EMBO J.*, 11, 57.
[108] Kadokura, H., Tian, H., Zander, T., Bardwell, J.C.A., Beckwith, J. 2004, *Science*, 303, 534.
[109] Missiakas, D., Georgopoulos, C., and Raina, S. 1993, *Proc. Natl. Acad. Sci. USA*, 90, 7084.
[110] Stafford, S.J., Humphreys, D.P., and Lund, P.A. 1999, *FEMS Microbiol. Lett.*, 174, 179.
[111] Metheringham, R., Griffiths, L., Crooke, H., Forsythe, S., and Cole, E. 1995, *Arch. Microbiol.*, 164, 301.
[112] Dailey, F.E., and Berg, H.C. 1993, *Proc. Natl. Acad. Sci. USA*, 90, 1043.
[113] Jacob-Dubuisson, F., Pinkner, J., Xu, Z., Striker, R., Padmanhaban, A., and Hultgren, S.J. 1994, *Proc. Natl. Acad. Sci. USA*, 91, 11552.
[114] Zapun, A., Bardwell, J.C.A., and Creighton, T.E. 1993, *Biochemistry*, 32, 5083.
[115] Wunderlich, M., Jaenicke, R., and Glockshuber, R. 1993, *J. Mol. Biol.*, 233, 559.
[116] Zapun, A., and Creighton, T.E. 1994, *Biochemistry*, 33, 5202.
[117] Nelson, J.W., and Creighton, T.E. 1994, *Biochemistry*, 33, 5974.
[118] Siedler, F., Rudolph-Bohner, S., Doi, M., Musiol, H.-J., and Moroder, L. 1993, *Biochemistry*, 33, 7488.

[119] Wunderlich, M., Otto, A., Seckler, R., and Glockshuber, R. 1993, *Biochemistry*, 32, 12251.
[120] Martin, J.L., Waksman, G., Bardwell, J.C.A., Beckwith, J., and Kuriyan, J. 1993, *J. Mol. Biol.*, 230, 1097.
[121] Guddat, L.W., Bardwell, J.C.A., and Martin, J.L. 1998, *Structure*, 6, 757.
[122] Graushopf, U., Winther, J., Korber, P., Zander, T., Dallinger, P., and Bardwell, J.C.A. 1995, *Cell*, 83, 947.
[123] Gilbert, H.F 1990, *Adv. Enzymol.*, 63, 69.
[124] Rivera-Madrid, R., Mesters, D., Marinho, P., Jacquot, J,-P., Decottiginies, P., Miginiac-Maslow, M.,and Meyer, Y. 1995, *Proc. Natl. Acad. Sci. USA*, 92, 5620.
[125] Bardwell, J.C.A., Lee, J.-O., Jander, G., Martin, N., Belin, D., and Beckwith, J. 1993, *Proc. Natl. Acad. Sci. USA*, 90, 1038.
[126] Tan, J., Lu, Y., and Bardwell, J.C.A. 2005, *J. Bacteriol.*, 187, 1504.
[127] Bader, M., Muse, W., Ballou, D.P., Gassner, CF., and Bardwell, J.C.A. 1999, *Cell*, 98, 217.
[128] Kishigami, S., and Ito, K. 1996, *Genes Cells*, 1, 201.
[129] Kishigami, S., Kanaya, E., Kikuchi, M., and Ito, K. 1995, *J. Biol. Chem.*, 270, 17072.
[130] Kobayashi, T., Takahashi, Y., and Ito, K., 2001, *Mol. Microbiol.*, 39, 158.
[131] Xie, T., Yu, L., Bader, M.W., Bardwell, J.C.A., and Yu, C.-A. 2002, *J. Biol. Chem.*, 277, 1649.
[132] Stewart, G.A., 1996, *Esherichia coli and Salmonella* – Cellular and Molecular Biology, American Society of Microbiology, Washington DC.
[133] Regeimbal, J., and Bardwell, J.C.A. 2002, *J. Biol. Chem.*, 277, 32706.
[134] Jander, G., Martin, N.L., and Beckwith, J. 1994, *EMBO J.*, 13, 5121.
[135] Inaba, K., Murakami, S., Suzuki, M., Nakagawa, A., Yamashita, E., Okada, K., and Ito, K. 2006, *Cell*, 127, 789.
[136] Guilhot, C., Jander, G., Martin, N., and Beckwith, J. 1995, *Proc. Natl. Acad. Sci. USA*, 92, 9895.
[137] Kobayashi, T., and Ito, K. 1999, *EMBO J.*, 18, 1192.
[138] Inaba, K., and Ito, K. 2002, *EMBO J.*, 21, 2646.
[139] Kadokura, H., Katzen, F., and Beckwith, J. 2003, *Ann. Rev. Biochem.*, 72, 111.
[140] Bader, M., Xie, T., Yu, C.-A., and Bardwell, J.C.A. 2000, *J. Biol. Chem.*, 275, 34, 26082.
[141] Inaba, K., Takahashi, Y.-h., and Ito, K. 2004, *J. Biol. Chem.*, 279, 6761.
[142] Kadokura, H., and Beckwith, J. 2002, *EMBO J.*, 21, 2354.
[143] Inaba, K., Takahashi, Y.-h., and Ito, K. 2005, *J. Biol. Chem.*, 280, 33035.

[144] Belin, P., Quemeneur, E., and Boquet, P.L. 1994, *Mol. Gen. Genet.*, 242, 23.
[145] Debarbieux, L., and Beckwith, J. 1999, *Cell*, 99, 117.
[146] Humphreys, D.P., Weir, N., Mountain, A., and Lund, P.A. 1995, *J. Biol. Chem.*, 270, 28210.
[147] Hinicker, A.,, and Bardwell, J.C.A. 2003, *Biochemistry*, 42, 1179.
[148] Missiakas, D., Georgopoulos, C., and Raina, S. 1994, *EMBO J.*, 13, 2013.
[149] Shevchik, V.E., Condemine, G., and Robert-Baudouy, J. 1994, *EMBO J.*, 13, 2007-2012.
[150] Zapun, A., Missiakas, D., Raina, S., and Creighton, T.E. 1995, *Biochemistry*, 34, 5075.
[151] Rietsch, A., Belin, D., Martin, N., and Beckwith, J. 1996, *Proc. Natl. Acad. Sci. USA*, 93, 13048.
[152] Sone, M., Akiyama, Y., and Ito, K. 1997, *J. Biol. Chem.*, 272, 10349.
[153] Joly, J.C., and Swartz, J.R. 1997, *Biochemistry*, 36, 10067.
[154] Rietsch, A., Bessette, P., Georgiou, G., and Beckwith, J. 1997, *J. Bacteriol.*, 179, 6602.
[155] Darby, N.J., Raina, S., and Creighton, T.E. 1998, *Biochemistry*, 37, 783.
[156] Maskos, K., Huber-Wunderlich, M., and Glockshuber, R. 2003, *J. Mol. Biol.*, 325, 495.
[157] Chen, J., Song, J.-l., Zhang, S., Wang, Y., Cui, D.-f., and Wang, C.-c. 1999, *J. Biol. Chem.*, 274, 19601.
[158] Zhao, Z., Peng, Y., Hao, S.-f., Zeng, Z.-h., and Wang, C.-c. 2003, *J. Biol. Chem.*, 278, 43292.
[159] Bader, M.W., Hiniker, A., Regeimbal, J., Goldstone, D., Haebel, P.W., Riemer, J., Metcalf, P., and Bardwell, J.C.A. 2001, *EMBO J.*, 20, 1555.
[160] Segatori, L., Murphy, L., Arredondo, S., Kadokura, H., Gilbert, H., Beckwith, J., and Georgiou, G. 2006, *J. Biol. Chem.*, 281, 4911.
[161] Andersen, C.L., Matthey,-Dupraz, A., Missiakas, D., and Raina, S. 1997, *Mol. Microbiol.*, 26, 121.
[162] Bessette, P.H., Cotto, J.J., Gilbert, H.F., and Georgiou, G. 1999, *J. Biol. Chem.*, 274, 7784.
[163] Shao, F., Bader, M.W., Jacob, U., and Bardwell, J. C.A. 2000, *J. Biol. Chem.*, 275, 13349.
[164] Van Straaten, M., Missiakas, D., Raina, S., and Darby, N.J. 1998, *FEBS Letters*, 428, 1998.
[165] Hinicker, A., and Bardwell, J.C.A. 2004, *J. Biol. Chem.*, 279, 12967.
[166] Heras, B., Edeling, M.A., Schirra, H., Raina, S., and Martin, J.L. 2004, *Proc. Natl. Acad. Sci. USA,* 101, 8876.

[167] Chung,J., Chen, T., and Missiakas, D. 2000, *Mol. Microbiol.*, 35, 1099.
[168] Crooke, H., and Cole, J. 1995, *Mol. Microbiol.*, 15, 1139.
[169] Missakas, D., Schwager, S., and Raina, S. 1995, *EMBO J.*, 14, 3415.
[170] Fong, S.T., Camakaris, J., and Lee, B.T. 1995, *Mol. Microbiol.*, 15, 1127.
[171] Goulding, C.W., Sawaya, M.R., Parseghian, A., Lim, V., Eisenberg, D., and Missiakas, D. 2002, *Biochemistry*, 41, 6920.
[172] Haebel, P., Goldstone, D., Katzen, F., Beckwith, J., and Metcalf, P. 2002, *EMBO J.*, 18, 4774.
[173] Engh, R.A., and a.H., R. 1991, *Acta Crystallogr.*, A 47, 392.
[174] Page, M.D., Saunders, N.F.W., and Ferguson, S.J. 1997, *Microbiology*, 143, 3111.
[175] Gordon, E.H., Page, M.D., Willis,A.C., and Ferguson, S.J. 2000,. *Mol. Microbiol.*, 35, 1360.
[176] Stewart, E.J., Katzen, F., and Beckwith, J., 1999, *EMBO J.*, 18, 5963.
[177] Katzen, F., and Beckwith, J. 2003, *Proc. Natl. Acad. Sci. USA*, 100, 10471.
[178] Katzen, E., and Beckwith, J. 2000, *Cell*, 103, 769.
[179] Kim, J.H., Kim, S.J., Jeong, D.G., Son, J. H.J., and Ryu, S.E. 2003, *FEBS Lett.*, 543, 164.
[180] Stirnimann, C.U., Rozhkova, A., Grauschopf, U., Bockmann, R.A., Glockshuber, R., Capitani, G., and Grutter, M.G.2006, *J. Mol. Biol.*, 358, 829.
[181] Rozhkova, , A., Stirnimann, C.U., Frei, P. Grauschopf, U., Grutter, M.G., Capitani, G., and Glockshuber, R. 2004, *EMBO J.*, 23, 1709.
[182] Collet, J.-F., Riemer, J., Bader, M.W., and Bardwell, J.C.A. 2002, *J. Biol. Chem.*, 277, 26886.
[183] Aslund, F., Berndt, K.D., and Holmgren, A. 1997, *J. Biol. Chem.*, 272, 30780.
[184] Krupp, R., Chan, C., and Missiakas, D. 2001, *J. Biol. Chem.*, 276, 3696.
[185] Russel, M. 1995, *Methods Enzymol.*, 252, 264.
[186] Goldstone, D., Haebel, P.W., Katzen, F., Bader, M.W., Bardwell, J.C.A., Beckwith, J., and Metcalf, P. 2001, *Proc. Natl. Acad. Sci. USA*, 98, 9551.

Index

A

acceptor, 15, 17, 20, 21, 24, 35, 37
acceptors, 21, 37, 52
access, 27, 52, 60
accessibility, 55
acid, 4, 5, 6, 8, 36, 42, 43
acidic, 5, 6, 31, 44
activation, 54, 55
active centers, 60
active site, vii, 4, 5, 6, 7, 8, 9, 11, 12, 13, 14, 15, 19, 21, 22, 23, 25, 30, 31, 32, 39, 40, 41, 42, 43, 46, 47, 48, 49, 50, 51, 52, 54, 55, 56, 57, 58, 59, 60
adenine, 20, 21
aerobic, 21, 34, 37, 38
agent, 11, 27, 39
aggregation, 41
aid, vii
air, 58
alanine, 15, 19, 52
alkaline, 42, 45, 47
alkaline phosphatase, 42, 45, 47
allele, 18
amino, 4, 5, 6, 8, 11, 18, 36, 42, 43
amino acid, 4, 5, 6, 8, 11, 18, 36, 42, 43
amino acids, 6, 11, 18, 43
amylase, 40
anaerobic, 21, 24, 37, 38
analytical techniques, 19
anion, 9, 31, 36
antigen, 47
arginine, 10, 11
aromatic, 46, 47
aspartate, 5
asymmetry, 23
atoms, 4, 7, 22, 46, 55
attacks, 12, 34, 35, 51
attention, 1, 60
availability, 2
avoidance, 43

B

bacteria, vii, 29, 33, 37, 44, 57, 60
bacterial, vii, 2, 29, 37, 41, 44, 47, 56, 59, 60
bacterial cells, 56, 60
bacterium, 31
behavior, 22
binding, 6, 8, 16, 20, 23, 33, 34, 40, 41, 43, 44, 47, 54, 55, 56
biochemical, 20, 49, 50, 59
biosynthesis, 25
birds, 11

blocks, 20
bonding, 22
bonds, vii, 1, 3, 4, 6, 9, 12, 13, 14, 15, 16, 17, 18, 19, 23, 25, 26, 27, 29, 30, 32, 33, 34, 36, 37, 38, 39, 40, 41, 42, 51, 53, 57, 59, 60
bovine, 8, 16, 41
buffer, 17, 25, 26, 27, 28, 29, 32, 39

C

capacity, 18
carrier, 37, 38, 59
catalysis, 1, 6, 8, 12, 15, 21, 23, 35
catalyst, vii, 1, 8, 12, 14, 15, 17
catalysts, 1
catalytic, 2, 6, 8, 10, 11, 16, 17, 22, 23, 35, 40, 41, 43, 47, 54, 55, 57
catalytic activity, 6, 35
catalytic properties, 22, 40
cDNA, 12
cell, 1, 6, 18, 26, 27, 30, 37, 40, 60
chaperones, 2
chemical, 6, 31
chemical properties, 31
chloride, 31
chromatography, 3
classical, 44
codes, 12
coding, 30
complement, 12, 18, 37, 42, 43, 57
complementary, 53
complexity, 15
components, 20, 35, 38, 45, 46, 50, 59
concentrates, 17
concentration, 3, 16, 25
configuration, 12
conformational, 4, 16, 31, 32, 41, 43, 48, 49, 53, 54, 60
Congress, iv
conservation, 18
constraints, 29
consumption, 20
controlled, 28, 43
conversion, 26
copper, 44

couples, 20
covalent, 13, 14, 40
credibility, 39
cross-linked, 1
cross-linking, 25
crystal, 2, 8, 9, 21, 22, 32, 33, 46, 48, 53, 54
crystal structure, 2, 8, 9, 21, 32, 33, 46, 48, 53, 54
crystallization, 48
crystallographic, 22
C-terminal, 3, 5, 7, 9, 10, 11, 13, 15, 18, 22, 23, 33, 41, 42, 43, 45, 46, 47, 53
C-terminus, 6, 9, 42, 53
cycles, 15, 16, 59
cysteine, vii, 4, 6, 7, 9, 10, 11, 12, 13, 15, 16, 19, 20, 22, 24, 25, 31, 33, 34, 39, 40, 47, 48, 49, 50, 51, 52, 54, 55, 57, 58, 59, 60
cysteine residues, vii, 4, 6, 7, 9, 10, 11, 15, 16, 19, 20, 24, 31, 34, 39, 40, 47, 49, 51, 54, 55, 57, 59, 60
cytochrome, 35, 37, 38, 44
cytochrome oxidase, 35, 37, 38
cytoplasm, vii, 29, 32, 33, 46, 49, 52, 53, 58, 60
cytoplasmic membrane, vii, 48, 59
cytosol, 3, 17, 27, 29, 47, 60
cytosolic, 25, 27, 60

D

de novo, 35, 40
defects, 20, 33
degree, 22, 25, 37
dehydrogenase, 41
denaturation, 31
denatured, 1, 41
density, 48
deposits, 20
derivatives, 49
deviation, 22
diffusion, 29
dimer, 16, 39, 41, 42, 43, 54, 56
dimeric, 41, 42, 43, 54, 59
dimerization, 15, 41, 42, 43, 59
Discovery, 30

displacement, 46
disulfide, vii, 1, 2, 3, 4, 6, 8, 9, 10, 11, 12, 13, 14, 15, 16, 17, 18, 19, 20, 21, 22, 23, 24, 25, 26, 27, 29, 30, 31, 32, 33, 34, 35, 36, 37, 38, 39, 40, 41, 42, 43, 44, 46, 47, 48, 49, 50, 51, 52, 53, 54, 55, 56, 57, 58, 59, 60
disulfide bonds, vii, 1, 3, 4, 6, 9, 12, 13, 14, 15, 16, 17, 18, 19, 23, 25, 26, 27, 29, 30, 32, 33, 34, 36, 37, 38, 40, 41, 42, 51, 57, 59, 60
disulfide isomerase, vii, 1, 2, 5, 15, 30, 39, 41, 42, 44, 54, 59, 60
divergence, 6
domain structure, 5, 11, 45
donor, 15, 32, 34, 59
donors, 52
DsbA, vii, 30, 31, 32, 33, 34, 35, 36, 37, 38, 39, 40, 41, 42, 43, 44, 47, 56, 57, 58, 59, 60
DsbB, vii, 33, 34, 35, 36, 37, 38, 39, 42, 43, 57, 58, 59
DsbC, vii, 15, 39, 40, 41, 42, 43, 44, 45, 49, 50, 51, 54, 55, 56, 57, 58, 59, 60
DsbD, vii, 44, 45, 46, 47, 48, 49, 50, 51, 52, 53, 54, 55, 57, 58, 59, 60
duplication, 6

E

E. coli, 6, 7, 30, 32, 39, 40, 44, 52, 57, 58, 59
electron, 17, 20, 21, 22, 24, 35, 36, 37, 48, 49, 50, 51, 52, 53, 54, 56, 57, 58, 59
electron density, 48
electronic, iv, 35, 36
electrons, 20, 21, 22, 24, 35, 37, 38, 45, 46, 47, 48, 49, 50, 51, 53, 56, 57, 58, 59, 60
electrostatic, iv, 9
encoding, 17, 25, 49
endoplasmic reticulum, vii, 1, 3, 4, 5, 17, 19, 20, 24, 25, 26, 27, 28, 29, 37, 57, 58, 59, 60
energy, 15, 31
energy transfer, 15
engineering, 5
envelope, 60
environment, 3, 29, 31, 32, 44, 46, 47, 59
enzymatic, 47

enzymatic activity, 47
enzyme, 4, 5, 7, 8, 9, 10, 11, 12, 13, 14, 20, 24, 25, 26, 30, 31, 35, 39, 41, 43, 44, 57
enzymes, vii, 9, 16, 29, 57, 60
equilibrium, 28, 31, 32
eukaryotes, vii, 20, 29, 38, 59
eukaryotic, vii, 1, 19, 20, 27, 37, 38, 57, 58, 59
eukaryotic cell, 20, 27, 37, 57, 58
evidence, 11, 12, 13, 20, 21, 22, 27, 31, 32, 33, 40, 48, 52
evolution, 6, 15
exclusion, 3
exogenous, 21
experimental condition, 50
expert, iv
exposure, 29
external environment, 29
extracellular, 29

F

FAD, 20, 21, 24
family, vii, 9, 18, 30, 37, 44, 59
flexibility, 22
flow, vii, 19, 22, 28, 36, 37, 38, 45, 50, 51, 58
fluorescence, 15
focusing, 2
foldases, vii, 2
folding, vii, 1, 4, 5, 8, 12, 13, 17, 19, 20, 24, 25, 26, 27, 29, 30, 31, 37, 40, 44, 45, 56, 57, 58, 59, 60
folding catalyst, 8
free energy, 31
fulfillment, 30
fumarate, 37
functional analysis, 27
fungi, 11

G

gene, 1, 6, 12, 17, 25, 30, 33, 39, 43, 44
genes, 57
genetic, 2, 17, 20, 30, 39, 45

glutamate, 5
glutathione, vii, 3, 17, 19, 25, 26, 27, 29, 32, 39, 40, 43, 58, 60
Gram-negative, vii, 29, 57
groups, 5, 9, 17, 30, 31, 39, 56
growth, 12, 18, 37

H

handling, 2
helix, 6, 7, 9, 10, 33, 43, 47, 48, 52
homogeneity, 50
homogenous, 13
homology, 6
human, 6, 11, 15, 16, 18
hybrid, 41, 47
hybridization, 42
hybrids, 42
hydrogen, 21, 41, 53
hydrogen bonds, 41, 53
hydrogen peroxide, 21
hydrophobic, 7, 8, 15, 33, 41, 44, 45, 47, 49, 53
hydrophobic interactions, 16
hypothesis, 13, 25

I

identification, 33
identity, 21, 43
immunoglobulin, 1, 30, 46, 47, 53, 56
in vitro, 1, 12, 20, 25, 26, 35, 39, 40, 50
in vivo, 1, 11, 19, 20, 25, 26, 27, 30, 39, 42, 44
inactivation, 9, 39
inactive, 6, 16, 41, 47
induction, 18
inhibitor, 8, 16, 27, 40, 41
inhibitory, 10
inhibitory effect, 10
initial state, 20
injury, iv
insight, 21, 51

interaction, 6, 8, 9, 15, 17, 23, 43, 44, 47, 49, 53, 54
interactions, 8, 16, 28, 31, 32, 40, 42, 46, 52, 53, 54, 55
interface, 8, 41, 53
intermolecular, 19, 35, 36, 52, 54, 55
internal clock, 13
intrinsic, 23
ionization, 9
isolation, 17
isomerization, vii, 1, 2, 12, 13, 14, 15, 16, 23, 25, 26, 27, 29, 38, 40, 42, 45, 51, 57, 58, 59, 60

K

kinetics, 26, 28, 40

L

labeling, 54
liberation, 36
ligand, 7
lipid, 29, 47
location, 7, 11, 48
London, 62
luciferase, 44
lumen, vii, 1, 3, 4, 17, 18, 27, 37, 59

M

machinery, 25
magnetic, iv
maintenance, 46
mammalian cell, 27
mammalian cells, 27
mammals, 3, 11
mechanical, iv
membranes, 60
metabolism, 37
microbial, 16
microsomes, 20, 27
mirror, 32
misfolded, vii, 8, 16, 40

misfolding, 39
model system, 23
models, 48
modules, 6
molecular mass, 3, 44
molecular oxygen, 20, 21
molecules, 16, 29, 54, 55
monomer, 3, 41, 42, 54, 55, 60
monomeric, 15, 30, 42, 54, 59
monomers, 15, 41, 56
multidimensional, 6
mutagenesis, 31, 32
mutant, 12, 13, 18, 19, 25, 32, 33, 42, 43, 45, 49, 52
mutant proteins, 12
mutants, 13, 18, 30, 32, 42, 43, 44, 54
mutation, 44
mutations, 33, 37

N

natural, 23
network, 32
New York, iii, iv
NMR, 6
non-native, vii, 4, 12, 14, 15, 16, 26, 27, 39, 40, 57, 59
normal, 13, 26, 31, 34, 49
normal conditions, 13
N-terminal, 5, 9, 11, 12, 13, 15, 22, 23, 25, 30, 31, 32, 33, 35, 41, 43, 45, 46, 49, 53
nucleophilicity, 9

O

observations, 21, 39, 40
organelle, 4
organism, 11
orientation, 7, 43
oxidation, vii, 10, 11, 12, 14, 15, 16, 17, 18, 19, 20, 21, 23, 25, 26, 27, 31, 33, 34, 35, 36, 37, 39, 42, 43, 48, 49, 53, 55, 57, 58, 59, 60
oxidation rate, 23

oxidative, vii, 1, 13, 17, 19, 20, 22, 24, 25, 26, 27, 29, 32, 37, 42, 59, 60
oxidative stress, 26, 27
oxygen, vii, 20, 21, 24, 35, 37, 38, 58, 60

P

pairing, 16
pancreatic, 8, 16, 41
PapD, 30
passive, 29
pathways, vii, 18, 27, 29, 37, 38, 42, 58, 60
PDI, v, vii, 1, 3, 4, 5, 6, 7, 8, 9, 10, 11, 12, 13, 14, 15, 16, 17, 18, 19, 20, 21, 22, 23, 24, 25, 26, 30, 37, 41, 57, 58, 59, 60
peptide, 7, 13, 15, 16, 25, 40, 41, 43
peptides, 7, 8, 31, 38, 39, 40, 50
periplasm, 2, 29, 30, 33, 37, 38, 39, 40, 42, 44, 45, 47, 49, 51, 52, 53, 57, 58, 59
permeant, 26
peroxide, 21
pH, 9, 29, 31, 32, 36, 49
phenotypes, 43
phosphate, 41
physiological, 9, 31, 39
planar, 53
Plasmids, 49
play, 9, 13, 15, 23, 26, 32, 48, 49
polypeptide, 4, 6, 15, 17, 50
polypeptides, 49, 50
poor, 15, 27
porous, 29
post-translational, 4, 5
power, 32, 35, 58
prediction, 33
preference, 22
preparation, iv
prokaryotes, 2, 35, 38, 59
prokaryotic, vii, 2, 37, 38, 59
prokaryotic cell, vii, 37
promote, 23
property, iv, 17, 33, 36
protection, 23, 25
protein, vii, 1, 2, 3, 4, 5, 6, 8, 11, 12, 13, 15, 16, 17, 18, 19, 20, 21, 22, 23, 24, 25, 26,

27, 30, 31, 32, 33, 37, 38, 39, 40, 41, 43, 44, 45, 49, 52, 53, 54, 56, 57, 58, 59, 60
protein crystallography, 49
protein disulfide isomerase, vii, 1, 15, 41, 54
protein engineering, 5
protein folding, 1, 8, 12, 17, 19, 20, 25, 26, 37, 45, 56, 60
protein oxidation, 18, 19, 20, 25, 37, 59
protein sequence, 18
protein-protein interactions, 28
proteins, iii, vii, 1, 2, 3, 6, 8, 12, 16, 17, 18, 19, 25, 27, 28, 29, 30, 32, 33, 37, 38, 39, 40, 41, 44, 45, 47, 49, 54, 56, 57, 58, 59, 60
proteolysis, 5, 41

relationship, 19, 21
research, 2, 20, 39, 43, 60
researchers, 60
residues, vii, 4, 5, 6, 7, 8, 9, 10, 11, 15, 18, 19, 20, 22, 24, 31, 32, 33, 34, 36, 39, 40, 41, 44, 46, 47, 48, 49, 51, 53, 54, 55, 56, 57, 59, 60
resistance, 43, 44
resolution, 32, 33, 34, 35, 36, 48, 49, 53, 60
retention, 5
reticulum, vii, 1, 3, 4, 5, 17, 19, 20, 24, 25, 26, 27, 28, 29, 37, 57, 58, 59, 60
riboflavin, 20
rigidity, 49, 53
rings, 21

Q

quinone, 33, 34, 57
quinones, 35

R

Raman, 62
random, 17, 56
range, 3, 13, 21, 27, 31, 32, 44
rat, 5
reaction center, 33
reaction rate, 31
reactivity, 9, 16, 28
reagents, 31
reception, 35
recognition, 3, 54, 56
recovery, 27
red shift, 36
redox, 3, 11, 17, 19, 20, 22, 25, 27, 28, 29, 30, 31, 32, 34, 36, 39, 40, 43, 49, 50, 51, 52, 54, 56, 60, 63
redox-active, 19, 20, 28, 60
reductases, 26
reduction, 11, 12, 13, 14, 15, 16, 17, 20, 21, 26, 27, 31, 32, 33, 36, 40, 45, 46, 48, 49, 53
regenerate, 14
regeneration, 26, 35, 36, 59
regulation, 25

S

Saccharomyces cerevisiae, 12
Salmonella, 66
search, 1, 10, 13
secretion, 3
selecting, 16
separation, 33, 42
series, 6, 48, 50
serine, 34
services, iv
shape, 3, 8, 15, 53
shares, 6
shuttles, 37
signs, 33
similarity, 15, 31, 32, 43, 44, 60
sites, 5, 8, 13, 16, 22, 23, 30, 32, 41, 42, 43, 54, 55, 58, 60
size-exclusion chromatography, 3
solvent, 48
spatial, 16
species, 33, 39
specificity, 44, 51
spectroscopy, 36
spectrum, 52
stability, 30, 31, 46, 47, 49
stabilization, 31
stabilize, 31
stages, 4, 5, 12

steric, 22
stiffness, 49
strain, 16, 18, 26, 30, 49
strains, 26, 30
stress, 26, 27
subdomains, 47
substitution, 52
substrates, 6, 8, 10, 13, 15, 17, 20, 25, 31, 35, 44, 50, 56, 60
sulfur, 7, 22, 47, 55
Sun, 63
supply, 39
surface area, 8
synthesis, 1, 44
systematic, 23, 34, 44
systems, 29, 38, 59

T

temperature, 18
temporal, 12
tension, 31
theoretical, 48
thermodynamic, 49
thermodynamic stability, 49
thioredoxin, vii, 5, 6, 7, 8, 9, 15, 23, 30, 32, 33, 37, 39, 41, 42, 43, 45, 46, 47, 48, 49, 50, 51, 52, 53, 54, 56, 57, 58, 59, 60
three-dimensional, 8, 31, 41, 42
time, 11, 13, 16, 20, 26, 34, 48, 50
Tokyo, 61
topological, 48
topology, 34, 47
transfer, 4, 15, 19, 20, 21, 25, 28, 35, 36, 37, 46, 48, 49, 50, 51, 52, 53, 56, 57, 59, 60
transition, 31, 36
transitions, 31
translation, 1
translational, 4, 5
translocation, 1, 47
transmembrane, 33, 45, 47, 48, 53
transport, 35, 48, 54, 57
trial, 16
trial and error, 16
triggers, 36
trypsin, 8, 16, 40, 41
tryptophan, 15
two-state model, 31

U

unfolded, 8, 15, 17, 18, 23, 29, 31
unfolded protein response, 18

V

validity, 54
values, 9, 10, 23, 31, 48
van der Waals, 46
variation, 56
visible, 36

W

Washington, 66
wild type, 32, 33, 40, 41, 43, 49
workers, 1, 13

X

X-ray, 23, 53

Y

yeast, 1, 3, 7, 11, 17, 18, 19, 20, 22, 24, 25
yield, 36